UAMR Studies on Development and Global Governance

Band 62

T0135492

UAMR Studies on Development and Global Governance

vormals Bochum Studies in International Development und
Bochumer Schriften zur Entwicklungsforschung
und Entwicklungspolitik

Band 62

The UMAR Graduate Centre for Development Studies is a collaboration project between the Institute of Development Research and Development Policy (IEE), Ruhr-University Bochum, the Institute of Political Science (IfP) and the Institute for Development and Peace (INEF), both located at the University Duisburg-Essen. The Centre is part of the University Alliance Metropolis Ruhr (UAMR) which aims at establishing the region as a cluster of excellence in research and training.

Herausgegeben für das UAMR Graduate Centre
for Development Studies von:

Prof. Dr. Matthias Busse, Prof. Dr. Tobias Debiel,
Prof. Dr. Christof Hartmann, Prof. Dr. Markus Kaltenborn,
Prof. Dr. Helmut Karl, Prof. Dr. Wilhelm Löwenstein

Abate Mekuriaw Bizuneh

Climate Variability and Change in the Rift Valley and Blue Nile Basin, Ethiopia

Local Knowledge, Impacts and Adaptation

Logos Verlag Berlin

λογος

UAMR Studies on Development and Global Governance

Herausgegeben von:

Prof. Dr. Matthias Busse, Prof. Dr. Tobias Debiel,
Prof. Dr. Christof Hartmann, Prof. Dr. Markus Kaltenborn,
Prof. Dr. Helmut Karl, Prof. Dr. Wilhelm Löwenstein

Institut für Entwicklungsforschung und Entwicklungspolitik
Ruhr-Universität Bochum
Universitätsstr. 150
D-44801 Bochum

Telefon: +49(0)234/32-22418
Telefax: +49(0)234/32-14294
E-mail: IEEOffice@ruhr-uni-bochum.de
http://www.uamr-graduate-centre.org

Bibliographic information published by the Deutsche Nationalbibliothek

The Deutsche Nationalbibliothek lists this publication in the Deutsche Nationalbibliografie; detailed bibliographic data are available in the Internet at http://dnb.d-nb.de.

Zugl.: Bochum, Univ., Dissertation 2013

© Copyright Logos Verlag Berlin GmbH 2013
Alle Rechte vorbehalten.

ISBN 978-3-8325-3524-7
ISSN 2194-167X

Logos Verlag Berlin GmbH
Comeniushof, Gubener Str. 47,
10243 Berlin
Tel.: +49 (0)30 / 42 85 10 90
Fax: +49 (0)30 / 42 85 10 92
http://www.logos-verlag.de

Dedication

To the person who gave her entire life to make sure that I and my siblings attend school - my heroine mother, Walubish Goadie.

Acknowledgement

Many thanks to give for the people who have helped me to complete my dissertation successfully. First, I would like to thank my supervisors Professor Dr. Bernhard Butzin and Professor Dr. Wilhelm Löwenstein for their priceless guidance and support. I feel privileged to have both of them as my supervisors and it has been an excellent opportunity and experience to learn from their insightful minds. Among the many things I benefited, Professor Dr. Bernhard Butzin's thoughts in looking into the research problem from different angles and unreserved help in offering detailed constructive comments, and Professor Dr. Wilhelm Löwenstein's tremendous encouragement and invaluable guidance particularly in the application of quantitative methods, has inspired me not only throughout the research process but the academic and professional world yet to come.

I would like to express my sincere gratitude to Catholic Academic Exchange Service (KAAD) for sponsoring my study in Germany. My deepest appreciation goes also to the Institute of Development Research and Development Policy (IEE) for providing additional financial support towards the completion of my PhD project. I gratefully acknowledge the Research School of Ruhr University Bochum for providing financial assistance to conduct the field work and giving me the chance to participate in several of skilled oriented seminars and trainings it offers. My gratitude also goes to the staff and colleagues at IEE who have been helpful and supportive throughout my stay at the Institute, and in this list I want to add Prof. Dr. Katja Bender, the ex-coordinator of the PhD IDS (International Development Studies) program for her keen assistantship in administrative and academic matters, and in particular mention Dr. Martina Shakya for her willingness to share her knowledge and experience in the topic of my research.

I am also indebted to Dr. Kindie Tesfaye for his encouragement, support and valuable hints starting from the beginning of my study. Dr. Ing. Henok Fikre also deserves my appreciation for proof reading some of the chapters of the dissertation. I am grateful for my brother Yaregal Mekuriaw and several friends who have been immensely helpful at the time of the field work. I would like to thank also many people and institutions that I cannot mention their names one by one; but to mention some are, Ethiopian Metrological Agency, Central Statistics Agency of Ethiopia, Health and Agricultural and Rural Development offices, and Agricultural Development Agents in the respective research sites for their cooperation in providing information. I also thank Dr. Gabriele Bäcker, Ruth Knoblich, Elkhan Sadik-Zada and Irene Wedler who helped me throughout the publication process.

Finally, those who are part of my life will not mind to appear in this last paragraph. In fact, I might not have paced to this last paragraph had it not been their unconditional support, motivation and love. Although my departure from them for the sake of the study was painful, they endured patiently. Even if shouldering all the responsibility at home and looking after the kids is beyond to bear, she did it with all heart and strength. Thank you for your patience and understanding and God richly bless you my wife, Lubaba, and my kids, Nuhamin and Kidanemariam. To God be the glory!

Table of Contents

Dedication i

Acknowledgement iii

Table of Contents v

List of Figures viii

List of Tables ix

List of Acronyms x

Chapter one 1

Introduction 1

1.1. Background 1

1.2. Statement of the Problem 4

1.3. Basic Questions 6

1.4. Objectives of the Study 7

1.5. Organization of the Study 7

Chapter Two 9

Climate Change and Variability: The Discourse, Measuring the Economic Impacts and Adaptation in Agriculture 9

Introduction 9

2.1. Climate Change Discourse 10

2.2. Economic Approaches to Measure the Impact of Climate on Agriculture 17

2.2.1. Structural Approaches 17

2.2.2. Spatial Analogue Approaches 19

2.2.3. The Production Function 26

2.3. Adaptation to Climate Variability and Change 29

2.3.1. Conceptualizing Adaptation 30

2.3.2. Adaptation to Climate Variability and Change in Agriculture 33

2.3.3. Theoretical Approaches to Agricultural Adaptation at Micro-Levels 35

Concluding Remark 42

Chapter Three 44

Theoretical and Methodological Framework 44

Introduction 44

3.1. Conceptualizing Climate Variability and Climate Change 44

3.2. Theoretical Framework 48
3.3. The Research Approach 53
3. 4. The Study Area and the Selection Process 55
3.5. Sampling and Data Collection Process 59
3.6. The Design of Questionnaire and Interview Procedure 61
3. 7. Data Analysis 62
Concluding Remark 63
Chapter Four 64
Exploring Local Knowledge on Climate Variability and Change 64
Introduction 64
4.1. Analytical Approach 64
4.2. Demographic Characteristics of the Respondents (Sample Households) 66
4.3. Farmers' Perception and the Trend of Climate Variables 69
4.4. Farmers' Description, Detection and Attribution of Climate Variability and Change 77
4.4.1. Detection through Farmers' Concrete Experiences 83
4.4.2. Farmers' Attribution of Observed Changes in the Local Climate 85
4.5. Farmers' Description of Impacts of Climate Variability and Change 90
4.6. Climate Stress Factors in the Face of Non-Climate Stressors 98
Concluding Remark 101
Chapter Five 103
The Economic Impact of Climate Variability and Change on Crop Production 103
Introduction 103
5.1. The Model Framework 103
5.2. Model Specification 108
5.2.1. Stage one 108
5.2.2. Stage two 109
5.3. Estimation Technique and Econometric Issues 110
5.4. Model Estimation and Discussion 113
5.5. Projecting the Impacts of Climate Change and Weighing the Impact of Climate Variability 117
Concluding Remark 122

Chapter Six **123**
Adaptive Behavior among Farmers **123**
Introduction 123
6.1. Model Framework 123
6.1.1. Framing Adaptive Behavior from Psychological Perspective 125
6.1.2. Framing Adaptive Behavior from SEI Perspective 127
6.2. Empirical Model 128
6.3. Assumptions and Model Performance in Logistic Regression 130
6.4. Descriptive Analysis 133
6.4.1. Farmers' Adaptive Strategies 133
6.4.2. Barriers to Adaptation 134
6.5. Regression Analysis and Discussion 137
6.5.1. Impact of Perception Factors on Adaptive Behavior 138
6.5.2. Impact of SEI Factors on Adaptive Behavior 141
Concluding Remark 146
Chapter Seven **148**
Conclusion and Outlook **148**
7.1. Revisiting the Research Problem and the Gaps in the Literature 148
7.2. Summary of Major Research Findings 149
7.3. Theoretical Contributions and Policy Implications of the Study and Outlook 152
7.3.1. Towards Local Discourse of Climate Variability and Change: Bridging the Gap in Climate Change Discourse 152
7.3.2. Towards the Very Form of Climate Critical to Subsistence Farming: Bridging the Gap in the Emphasis of Climate Concern 155
7.3.3. Towards Integrated Approach to Adaptive Behavior: Bridging the Gap between Psycho-cognitive and Socioeconomic Approaches 156
7.3.4. Towards Mitigation Actions at Local Levels: Bridging the Gap in Local Response 157
7.3.5. Towards Addressing the Needs of the Disadvantaged Groups: Bridging the Gap in Social Networks and Institutional Arrangements 158
References **159**
Appendix **182**

List of Figures

Figure 1.1. The relationship between national GDP growth, agricultural GDP growth and annual rainfall ... 3

Figure 3.1. Conceptual Framework (Key elements of the research are in italics) ... 52

Figure 3.2. Geographic location of the study areas ... 58

Figure 4.1. The major theme and issues addressed in this chapter (key elements are highlighted) ... 65

Figure 4.2. Age of heads of the households in categories ... 67

Figure 4.3. Education level of heads of the households ... 68

Figure 4.4. Farming experience of the household heads in years ... 68

Figure 4.5. Perception of farming households about the trend of precipitation in the last twenty years ... 70

Figure 4.6. Perception of farming households about the trend of temperature in the last twenty years ... 71

Figure 4.7. Trend of Precipitation and Temperature in Adet ... 72

Figure 4.8. Trend of Precipitation and Temperature in Wereta ... 73

Figure 4.9. Trend of Precipitation and Temperature in Alaba Kulito ... 73

Figure 4.10. Trend of Precipitation and Temperature in Arsi Negele ... 74

Figure 4.11. Trend of Precipitation and Temperature in Shashemene ... 74

Figure 4.12. Some of the houses damaged by the flood of April 2010 in Senbete (Shala Wereda) ... 80

Figure 4.13. A mass of land washed away by the flood of April 2010 in Senbete (Shala Wereda) ... 81

Figure 4.14. Land coverage by crops ... 87

Figure 4.15. Farmers' and experts' description and attribution of changes in local climate ... 89

Figure 4.16. Household members infected by malaria in percentage in Yilmana Densa (Adet Zuria and Mosebo Kebele) ... 93

Figure 5.1. The major themes and issues addressed in this chapter (key elements are highlighted) (Theme 2a and 2b). ... 104

Figure 6.1. The major themes and issues addressed in this chapter (key elements are highlighted) (Theme 3a and 3b) ... 124

Figure 6.2. Barriers to adaptation as identified by farmers ... 135

List of Tables

Table 2.1. Forms of Adaptation with examples 33
Table 3.1. The relationship between climate change and variability 47
Table 4.1. Sex of heads of the households 66
Table 4.2. Analysis of precipitation data from 1990 to 2009 75
Table 4.3. Analysis of temperature data from 1990 to 2009 76
Table 4.4. Longevity of food and beverages' fermentation and preservation in Blue Nile Basin Villages 84
Table 4.5. Longevity of food and beverages' fermentation and preservation in Rift Valley Lakes Basin Villages 84
Table 4.6. Summary of impacts of climate variability and change as experienced by farmers 98
Table 4.7. The Mean value of stress factors in descending order 100
Table 5.1. OLS estimation of the impact of conventional inputs on annual income from cereal crops 114
Table 5.2. OLS estimation of the impact of climate variables 116
Table 5.3. Projected income change in the mid-century 119
Table 5.4. OLS estimation of the impact of climate variability 121
Table 6.1. Adaptive strategies carried out by the households 133
Table 6.2. Logistic regression results with perceptual variables 139
Table 6.3. Logistic regression results with socioeconomic variables 143
Table 7.1. Comparing climate change discourse (global) and local discourse 154

List of Acronyms

AEZ	Agroecological Zone
AR5	Fifth Assessment Report
BLUE	Best Linear Unbiased Estimator
BNB	Blue Nile Basin
CGCM3.1	Coupled General Circulation Model, Version 3.1
CRU	Climate Research Unit
CSA	Central Statistics Agency
EDRI	Ethiopian Development Research Institute
EM-DAT	Emergency Events Database
EPA	Environmental Protection Authority
ERM	Environmental Resource Management
ESPERE	Environmental Science Published for Everybody Round the Earth
FAO	Food and Agriculture Organization
FGLS	Feasible Generalized Least Square
GCM	General Circulation Models
GDP	Gross Domestic Product
GFDL-CM2.0	Geophysical Fluid Dynamics Laboratory Climate Model, Version 2.0
GHG	Green House Gases
HDI	Human Development Index
IBP	International Business Publications
IFPRI	International Food Policy Research Institute
IGAD-ICPAC	Intergovernmental Authority on Development - Climate Prediction and Applications Centre
IPCC	Intergovernmental Panel on Climate Change
ITN	Insecticide-treated Nets
ML	Maximum Likelihood
MOFED	Ministry of Finance and Economic Development
MOH	Ministry of Health
MPPACC	Model of private proactive adaptation to climate change
MRI-CGCM2.3.2	Meteorological Research Institute - Coupled General Circulation Model, Version 2.3.2
NAS	National Academy of Sciences
NMA	National Meteorological Agency
NMSA	National Meteorological Services Agency
OLS	Ordinary Least Square
PMT	Protection Motivation Theory
RESET	Regression Specification Error Test
RVLB	Rift Valley Lakes Basin
SEI	Socioeconomic and Institutional
SRES	Special Report on Emissions Scenarios

TFP	Total Factor Productivity
TLU	Tropical Livestock Unit
UHI	Urban Heat Island effect
UNCTAD	United Nations Conference on Trade and Development
UNDP	United Nations Development Programme
UNEP	United Nations Environment Programme
UNFCCC	United Nations Framework Convention on Climate Change
USAID	United States Agency for International Development
VIF	Variance Inflation Factor
WMO	World Meteorological Organization

Chapter one

Introduction

1.1. Background

Climate plays an important role in shaping human and ecological systems. Human activities are naturally tied with climate and a change or variation in the climate directly or indirectly affects these activities. Particularly among nations and communities where economies are tied with primary sectors such as agriculture, climate remains a single important natural factor for the healthy functioning of the economies. For such economies, climate variability and change are major sources of concern. Among the economies of such concern, Ethiopia serves as a good example.

Ethiopia is well known for its confrontation and severe experience with climate related impacts. Its geographic location, low adaptive capacity and overdependence on rain-fed agriculture make the country one of the most vulnerable to climate variability and change. Geographically, Ethiopia is located in the eastern part of Africa which has been repeatedly attacked by climate related disasters. The country has been suffering from such disasters which manifest in the form of drought, flood, heavy rains, strong wind, high temperature and frost (Tadege, 2007). Of these hazards, drought and flood are the most important events severely damaging the economy and claiming the lives of people. Given that climate change likely increases the frequency and intensity of extreme events such as drought and flood (IPCC, 2007c), the occurrence of these adverse effects in the country are seemingly increasing from year to year.

In fact, Ethiopia has a long history of drought and the probability of recurrence of drought in the past was once in an interval of a decade. In recent decades, however, the frequency and extent appear to be growing (NMSA, 2001). For instance, the country experienced recurrent droughts in the years 1964-65, 1973/74, 1983/84, 1987/88, 1990-92, 1993/94, 2000, 2002/3, 2006, 2009 and recently in 2011 (Tadege, 2007[1]). Similarly, flood attacks occurred in 1988, 1993, 1994, 1995, 1996 and 2006 (Ibid). Following such recurrence of drought and the catastrophic drought of 1984-85, some climate researchers go further to believe Ethiopia as one of the first victims of climate change (Philander, 2008).

With regard to adaptive capacity, the country is endowed with little choice to respond to the impacts of climate variability and change. A simple indication that Ethiopia has

---

[1] Cited Quinn and Neal (1987); Degefu (1987); Nicholls (1993); Webb and Braun (1994) and others for the drought years up to 2002/2003.

low levels of adaptive capacity could be reflected by the country's low level of social and economic development by all standards (MOFED, 2003). In terms of HDI, Ethiopia ranked 174th in 2011 out of 187 nations of the world (UNDP, 2011); its per capita income is US$ 390 in 2010 which makes it much lower than the Sub-Saharan Africa average of US$ 1,165; and it still remains as one of the poorest countries with 38.9 percent of the population living under poverty line[2] in 2005 (http://data.worldbank.org/country/ethiopia[3]).

Agriculture is the main stay of the country's economy. It accounts about 43 percent of the GDP and 90 percent of the exports (Chanyalew et al., 2010:3), and serves as the direct source of employment and livelihood for about 85 percent of the population (NMSA, 2001:29). Out of the total amount of agricultural GDP, 95 percent comes from smallholder agricultural units (subsistence farming) (Chanyalew et al., 2010:3). Smallholder agriculture in the country is mainly constrained by old agricultural practices and structural problems, and hence it has fallen short of providing moderate and persistent income to farmers engaged in it (MOFED, 2003).

Agriculture's heavy dependence on rainfall is a major characteristic of the sector. Dependence on rainfall means that agricultural production is vulnerable to climatic variability and change, which in turn can severely affect food security and GDP growth. It follows that if the main source of an economy is rooted on climate sensitive sectors such as rain-fed agriculture, an episode of a single climate event could languish or even reverse the economic growth achieved in the past (IGAD-ICPAC, 2007). This is typically the case of Ethiopia where GDP growth is closely tied with rainfall - when the country gets adequate amount of annual rainfall, the GDP goes up, and when there is shortage of precipitation, the GDP strikingly goes down. This could be asserted easily by Figure 1.1 below which clearly shows the nexus between rainfall, on the one hand, and GDP and agricultural production, on the other hand. Similarly, Dercon's (2004) study of household consumption in rural Ethiopia shows that rainfall shock in one year has a lingering effect on households' welfare for many years to come. In his study, he indicated that a 10 percent rainfall decrease in one year has an impact of 1 percent point on the growth rates of about 4 to 5 years to come.

[2] National poverty line which is calculated based on population-weighted subgroup estimates from household surveys.

[3] Accessed 31/05/2012.

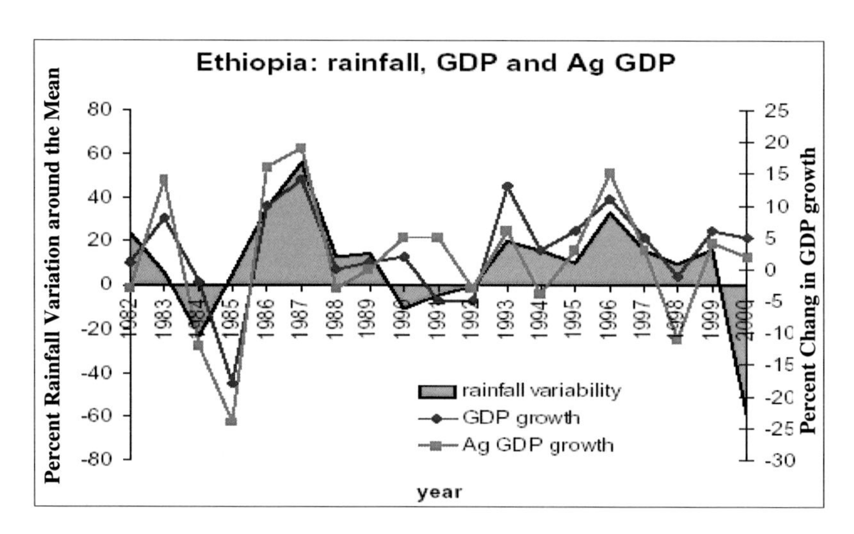

Figure 1.1 The relationship between national GDP growth, agricultural GDP growth and annual rainfall

Source: World Bank (2005) citing de Jong (2005), in IGAD-ICPAC (2007:32)

Such prone nature of the country's agriculture to rainfall indicates how climate variability and change could affect productivity. In the first place, agricultural productivity is lingering under stress from severe soil erosion, land degradation, deforestation, overgrazing and population pressure. Climate variability and change further exacerbate these conditions. Consequently, food requirement of the country remains at the mercy of climate. That is why understanding climate pattern vis-à-vis agricultural production in Ethiopia seems to be critical.

Studies on the trend of climate show that temperature has been increasing throughout the country with a mixed trend of precipitation. In the periods between 1953 and 1998, average annual rainfall showed a decreasing trend over the Northern half and Southwestern parts, and an increasing trend in the central part of the country. Average annual maximum temperature and average annual minimum temperature over the country have increased by 0.1°C and 0.25°C per decade, respectively (NMSA, 2001:73). As regard to the future, GCM (General Circulation Models) predictions also show generally an increasing trend of temperature with moderate inter-model differences. However, precipitation predictions are inconclusive, some showing negative trends and others positive trends (Hulme et al., 2001; Camberlin, 2009). Considering A1B emission scenario, for instance, mean annual temperature will increase in the range of 0.9 and 1.1 °C by the year 2030 and in between 1.7 to 2.1 °C by the year 2050 from the average of

1961 - 1990 (Tadege, 2007:23; IGAD-ICPAC, 2007:34). Whereas, the corresponding results for annual precipitation show a change in between 0.6 and 4.9 percent for 2030 and 1.1 to 18.2 percent for 2050 (IGAD-ICPAC, 2007:34).

Given that agricultural production is largely the reflection of climate and major natural resources (water, forest, energy, biodiversity etc) and the socio-economic development of Ethiopia is very much influenced by climate and its variability (NMSA, 2001), it is evident that climate variability and change are serious concerns to Ethiopia. Although drought is not a new phenomenon for Ethiopia, the frequency of its occurrence and the overall trend and predictions of climate over the country are becoming an ever increasing concern for food production. Consequently, how warming trends, changing precipitation and increasing climate variability (including droughts and floods) impact the country's agriculture and overall economy, along with the quest for adaptation, is a research area surging in this era of climate change discourse.

1.2. Statement of the Problem

It is widely acknowledged that poor and indigenous communities are among the most vulnerable and the first to confront the effects of climate events due to their reliance on natural resources (IPCC, 2007b; Salick & Byg, 2007). Nevertheless, the pendulum of policy and scientific discussion on climate variability and change has been mainly swinging among scientists and government delegates at higher levels. Besides, these discussions are largely drenched with controversy, skepticism and inaction (Cobb, 2011). Such state of the discourse has two clear implications to the vulnerable people around the developing world. First, since the discourse is mainly apprehended with lack of consensus (on the science, actions of response and funding mechanisms), world collective actions are more likely to delay for quite some time to come, and hence local communities (such as farming communities who rely on natural resources) will continue to suffer from the impacts. Secondly, due to its level, climate change discourse generally remains very weak at engaging local knowledge in ongoing scientific discussions and decision making. As such policies that eventually emanate from the discourse might be less responsive to local climate problems and specific vulnerabilities both in terms of policy ingredients and time frame. Consequently, alerting researchers and policy makers to look into alternative knowledge systems, such as local knowledge, to understand the impacts of climate variability and change at local levels so as to design response actions appropriate to the level.

Meanwhile, farming communities in the developing world largely remain unaware of ongoing scientific and policy debates on the global stage, and hence they live largely detached from the scientific understanding (knowledge) of climate variability and change. However, they have their own ways of understanding the climate, detecting, attributing and interpreting the changes based on experiential learning and indigenous knowledge passed to them through generations. It is through these mechanisms that they comprehend the impacts and inform their decision making towards adaptation. Therefore, understanding farmers' knowledge of climate variability and change should certainly form the foundation in research activity and policy formulation to address climatic concerns at local levels where scientific guidance and external support is minimal.

It has been long adhered that local (indigenous) knowledge, although underutilized, provides problem-solving strategies and decision making platforms in solving environmental and development problems among the poor and the vulnerable (World Bank, 1998; Nyong et al, 2007). In the context of climate variability and change, studies that engage local knowledge are scarce at the outset and are only emerging in recent years. Some of the studies stress the importance of incorporating local knowledge into scientific inquiry and formal mitigation and adaptation strategies (Nyong et al, 2007; Macchi et al., 2008; Kelman & West, 2009; Green & Raygorodetsky, 2010; Cobb, 2011; Raygorodetsky, 2011). But, much work remains to be done as regard to tracking, documenting, and systematically integrating local knowledge with scientific knowledge (Gagnon & Berteaux, 2009). Indeed, local knowledge is crucial in developing regional climate adaptation policies and to fill a paucity of scientific data (Green & Raygorodetsky, 2010) and supplement scientific knowledge. Particularly, local knowledge is crucial to address local vulnerabilities in this divided world on the issue of climate change and response actions. It also helps decision makers and major actors at local levels to design practice based coping strategies to minimize local effects of climate variability and change.

Other studies, on the other hand, are limited to superficial examination of perception (signal detection) of farmers as to how climate variables (most importantly precipitation and temperature) in the past several decades had been behaving (Maddison, 2006; Gbetibou, 2009; Hassen & Nhemachena, 2008; Deressa et al, 2009). These studies principally attempt to explain adaptive behavior through the premise that adaptation is a two step process where farmers perceive climate change and that perception leads to adaptation (adaptive behavior). With this assumption they try to identify socioeconomic and institutional resources that facilitate adaptation. Such basement in socioeconomic and institutional resources is primarily borrowed from adoption theories of agricultural innovations. The importance of resources in facilitating and influencing adaptive behavior is beyond doubt. Nevertheless, the existence of these factors per se might not be sufficient for adaptation. How perceived signal is interpreted by farmers themselves is rather important than the signal itself. The extent of perceived vulnerability and severity of the signal for farming are obviously crucial to initiate adaptation. These factors coupled with the perception of self efficacy to undertake adaptation strategies and the efficacy of adaptation strategies to cope with the perceived signals of the climate could play an important role. However, the process of adaptation from such perspective has been ignored and yet needs to be explored.

Besides understanding adaptive behavior, studies also attempt to measure the economic impact of climate variability and change on agriculture. Towards this end, several studies have been applying the Ricardian model quite extensively both in the context of developed and developing countries, and commercial as well as subsistence agriculture[4]. The model was developed in the 1990s and it attempts to quantify the impacts with the assumption that the value of a farmland under properly functioning markets is the reflection of climate and other environmental factors (Medelsohn et al., 1994). However, due to its assumptions that do not fit into the developing world and its implicit assumptions of perfect adaptation and other theoretical limitations, the applicability of the Ricardian approach mainly remains to be problematic. Therefore, the application of the approach needs to be reconsidered particularly in the context of subsistence agriculture.

1.3. Basic Questions

This study is organized in three interrelated themes. Each theme serves as the main question that the study strives to address with several sub-questions under each theme.

Theme 1: Exploring farmers' perception and experience of detection and attribution of climate variability and change along with impacts.

1.1. How similar/different are farmers' perception of climate variability and change with respect to scientific data (meteorological recordings) on the major parameters of climate?
1.2. How do farming communities describe ongoing climate trends and changes in their areas? Does their description correspond with experts' views?
1.3. To what factors do farmers attribute the changes (if any) in local climate?
1.4. What impacts have farmers observed due to changes in the local climate?
1.5. What factors among climatic and non-climatic stressors (risk or uncertainties) are more important sources of concern for farming households?

Theme 2: Analyzing the impacts of climate variability and change and adaptation on crop production.

2.1. What is the impact of climate variability and change on crop production?
2.2. What is the impact of adaptation on crop production?

[4] Refer chapter 3 for the characteristics.

Theme 3: Investigating the socioeconomic and psychosocial factors that influence adaptive behavior.
  3.1. What factor influence farmers' adaptive behavior to climate variability and change?
    3.1.1. What is the effect of perception (psychological) factors on adaptive behavior?
    3.1.2. What is the effect of socioeconomic and institutional variables on adaptive behavior?

1.4. Objectives of the Study

The general objective of the study is to analyze local knowledge, quantify the economic impact of climate variability and change on crop production and critically study how adaptation is governed at micro-level. To achieve this general objective, the study pursues the following specific objectives.

1. To examine the trend and characteristics of local climate in the last several decades based on farmers' knowledge, expert views and meteorological recordings.
2. To investigate farmers' knowhow in detecting and attributing climate variability and change.
3. To compare and contrast farmers' accounts of climate variability and change with that of experts (and also with meteorological recordings).
4. To assess the socioeconomic impacts of climate variability and change based on farmers' experience and perception.
5. To measure the impact of climate variability and change and associated adaptation on crop production.
6. To investigate under what conditions adaptation takes place at farmers' level.
7. To propose a simple approach that could be used in distinguishing climate-induced adaptive behavior from that of non-climate-induced behavior.

It is believed that the attainment of the intended objectives will inform policy decision making at local levels. Besides, the study contributes to theoretical and methodological approaches discussed at the closing chapter of this dissertation.

1.5. Organization of the Study

This dissertation is organized into seven chapters. As described above, it begins its presentation by highlighting the main research venture through the background of the research, statement of the problem and research questions. In chapter 2, it presents the review of literature on climate change discourse, measuring the impacts of climate on agriculture, and adaptation. The chapter finally outlines the gaps that this research strives to fill in.

Chapter 3 then follows by detailing the methodological and theoretical framework employed to address the research questions of the study. The next three chapters (chapters 4 - 6) are then devoted to the analysis and discussion of the results of the study.

The empirical part of the study begins in chapter 4 where it intensively investigates local knowledge on climate variability and change. Taking the analogy of the scientific steps involved in the study of climate change, it systematically presents farmers' understanding, detection, attribution and impacts assessment of climate variability and change. Chapter 5 entirely focuses on analyzing the impacts by taking crop production (main source of farmers' livelihoods) as a case to look into the problem in detail through quantitative methodological applications. Chapter 6, which is devoted to the study of adaptation, approaches the problem of adaptive behavior from two theoretical perspectives and discusses regression results from these two perspectives. Finally, chapter 7 presents the summary of major findings, conclusion, and policy and theoretical implications of the study along with the outlook.

Chapter Two

Climate Change and Variability: The Discourse, Measuring the Economic Impacts and Adaptation in Agriculture

Introduction

Climate change is regarded as one of the most serious challenges of the 21st century. The challenge is not only linked to its effects and impacts, but also of lack of consensus as to whether it is the product of human activities or natural processes. To shade light on this issue as a starting point, this chapter begins with discussing the discourse of climate change. It explains how the scientific community is stranded in polarized ideas as to whether climate is reality or not, and an anthropogenic or a natural cause. Out of this discussion, it finally attempts to illustrate the imperativeness of local knowledge to respond to the impacts at local levels while the discourse continues to stay stranded in various opposing ideas and controversies.

Following this discussion, it goes to addressing specifically how impact of climate change and variability is measured in the context of agricultural production. It discusses the most widely applied models vis-à-vis the possibility of applicability in the context of subsistence farmers in the developing world, and if not lay the foundation as to why they fall short of applicability.

The chapter also provides discussions on adaptation as one of the two policy options (besides mitigation) in responding to climate change and variability. In fact, adaptation has existed in the history of human's struggle to adapt with changing environment. As such, the term adaptation is not a new phenomenon until this era when the concern on rapidly changing and increasingly variable climate embarked as a new policy issue. Due to this, research on adaptation has been one of the priority areas in the field both at macro- and micro-levels. At macro-level, adaptation research addresses agricultural adjustments at national and regional levels. At micro-levels, its focus lays mainly on understanding the impacts of and the processes that influence adaptation (adaptive behavior) at farm level. The chapter pays attention to this issue and deals with the theoretical approaches that have been applied to understand adaptive behavior of farmers. Finally, it attempts to show the gaps that this research pursues to bridge in the existing theoretical and methodological frameworks.

2.1. Climate Change Discourse

Scientific attempt to identify changes in the earth's climate dates back to the nineteenth century (Richer, 2010). A foundation work by the Swedish scientist and Noble laureate Svente Arrhenius in 1896 remarks the beginning of the concept of greenhouse effects (Bodansky, 2001; Bolin, 2007; Philander, 2008; Richer, 2010; NAS, 2010). Although he demonstrated the contribution of carbon dioxide to global warming[5], it was yet in the 1950s that climate change began to emerge as a concern (Richer, 2010).

In the late 1950s, two prominent American researchers, Roger Revelle and Charles Keeling, came up with undisputed results which alerted the world (Philander, 2008; Bodansky, 2001). Revelle's findings revealed the amount of carbon dioxide absorbed by oceans to be only one-tenth the rate that scientists had believed and this witnessed the longer stay of carbon dioxide in the atmosphere (Fleming, 1998; Richer, 2010). Similarly, Keeling developed a new method for measuring the concentration of carbon dioxide in the air and had shown the rising trend of carbon dioxide in the atmosphere along with a rise in temperature (Bolin, 2007; Philander, 2008). These findings coupled with the earlier studies triggered concerns in the climate change discourse that were raised and discussed fiercely in the decades there after.

Research in the 1970s and 1980s remarked the application of sophisticated computer models and scientific experiments which in turn enhanced scientific confidence in the global warming studies (Bodansky, 2001). Following the review of various models, the US National Academy of Sciences in its report of 1979 supported the idea of global warming as an effect to carbon dioxide emission (Bodansky, 2001; Philander, 2008). Moreover, the recognition of other greenhouse gases such as methane and nitrous oxides in the mid 1980s strengthened the scientific hypothesis of global warming (Bodansky, 2001). Gradually, scientific discussions led to the politics of climate change which in turn triggered intergovernmental discussions. In 1979, the World Meteorological Organization (WMO) organized the first world climate change conference in Geneva (Bolin, 2007; Pittock, 2009; Richer, 2010) and the conference concluded by expressing its concern over gradual warming, and appealing nations to work towards preventing human induced climate change (Philander, 2008; Pittock,

---

[5] Global warming, often used by the media to describe climate change or usually mixed with climate change, refers to a rise in global surface temperature due to human induced greenhouse gases/anthropogenic causes (Giddens, 2008; Moffatt, 2004; Wallington, et al, 2009; http://agroclimate.org/climate_change, accessed 20/05/2012). It induces the climate to change (Majumder et al., 2010). Whereas, climate change is more than global warming and includes other variables of climate in addition to temperature. Since, the scientific understanding of the term climate change began with anthropogenic interference with the atmosphere and much of the debate is on the role of anthropogenic activities, the term global warming in this study is treated mainly to address the historical development of the science and associated debates with it.

2009). Out of this concern which was enriched in the successive international meetings, the Intergovernmental Panel on Climate Change (IPCC)[6], with the aim of providing a comprehensive assessment of the science of climate change, was established in 1988 by the joint effort of the World Meteorological Organization and the United Nations Environment Programme (UNEP) (NAS, 2010).

Two years later, IPCC produced its first assessment report which indicated rapid changing trends of the climate and urged the need for counteraction (Philander, 2008). This outcome of IPCC had served as the basis for discussion at the Earth Summit held in Rio de Janeiro in 1992 (Philander, 2008; Richer, 2010). It was in this summit that nations agreed on the problem and signed the United Nations Framework Convention on Climate Change (UNFCCC)[7] (UNFCCC, 1992; Böhringer & Finus, 2005; Richer, 2010). The convention stressed the need for serious actions to reduce the emission of anthropogenic greenhouse gases, and the Framework Convention was ratified in 1994 (UNFCCC, 1992; Depledge, 2004). Once the Framework was ratified, the next negotiations to date have been entirely dominated by Kyoto Protocol[8] and the fate of Kyoto Protocol in post-Kyoto period when it expires in 2012 (Bodansky, 2001).

Even though, the issue of climate change was able to transform from scientific into a policy dialogue as of the mid and the late 1980s (Bodansky, 2001), the discourse still faces two major problems: lack of consensus on the science (mainly on the cause of

---

[6] Intergovernmental Panel on Climate Change (IPCC) is a key UN scientific body established jointly by the World Meteorological Organization (WMO) and the United Nations Environment Programme (UNEP) in 1988 to provide objective information about climate change to decision-makers and others interested in climate change. It does not sponsor climate research itself but organizes research undertakings carried out by scientists over the world, and assesses scientific information related to climate change, evaluates environmental and socioeconomic impacts of climate change and formulates response strategies (IPCC, 2007a; Bolin, 2007; NAS, 2010).

[7] The United Nations Convention on Climate Change (UNFCCC) is an international treaty agreed in Rio de Janeiro in 1992 at a summit commonly known as the Rio Earth Summit, and was entered into force on 21 March 1994. The Convention acknowledges the danger of climate change and aims to stabilize greenhouse gas concentration in the atmosphere at a level that would prevent dangerous anthropogenic interference with the climate system. It puts greater responsibility on developed countries, as they bear the burden of historic responsibility of greenhouse gas emissions. Hence, apart from targeting to cut their GHG emissions, developed countries provide financial and technological support to developing countries for their actions to tackle climate change. The Convention is serviced by a Secretariat based in Bonn, Germany, and is administered under the United Nations Rules and Regulations (UNFCCC, 1992; www.UNFCCC.org, accessed 26/01/2012).

[8] The Kyoto Protocol is the first international agreement adopted in Kyoto, Japan, in December 1997 and entered into force on 16 February 2005. It is the first step to operationalize UNFCCC. It sets binding targets for developed countries to reduce a combined greenhouse gas emission of 5% during 2008 - 2012 with respect to the 1990 levels (UNFCCC, 1998; Böhringer & Finus, 2005). It also stipulates the establishment of Adaptation Fund to finance adaptation projects and programmes in developing countries which are Parties of the Protocol (Böhringer & Finus, 2005; UNFCCC, 2008).

climate change and its consequences) and on the action of response. When the issue of climate change was first synthesized by Arrhenius, there was very little attention due to the probable belief that change was too small (Richer, 2010). Later when research goes growing and anthropogenic prepositions began mounting, skeptics (or contrarians) emerge to question the role of anthropogenic causes.

Some skeptics deny the role of human activities in climate change and attribute the observed changes to natural processes; and hence refute actions of response. On the other hand, other groups of skeptics believe in human induced climate change but they conversely undermine the consequences (Giddens, 2008). According to this later group of skeptics, Urban Heat Island effect[9] (UHI) and water vapor are important man-made causes (CLI, 2009; Neu, 2009; Irwin & Williams, 2010) than carbon dioxide and thus cutting carbon dioxide emission cannot be a solution. Still others who acknowledge earth's warming, such as very known contrarian, Bjørn Lomborg, downplay the significance of anthropogenic effect (Philander, 2008; Pittock, 2009), and believe that other world problems[10] are more pressing than climate change (Lomborg, 2006; Giddens, 2008). According to Lomborg, the total impacts of climate change will be negligible and will not pose a devastating problem for the future and thus should be easily dealt with no urgency to emission reductions, which could otherwise be very expensive and might even result worse consequences than the original affliction (Lomborg, 2006; Pittock, 2009). However, Pittock (2009:71) debunks Lomborg's argument by stating that:

> *"His claims are value-judgments based on discounting the more severe possible impacts, technological optimism regarding our adaptive capacity, and technological pessimism regarding our ability to reduce greenhouse gas emissions at low cost."*

On the other hand, believers of climate change assert their position by drawing several evidences. Scientists in this group do believe the importance of natural processes but regard human activity as the major cause for the rapid warming in the last several decades and warn dire consequences unless timely solution is undertaken (IPCC, 2007c; Pittock, 2009). For them, carbon dioxide is the main contributor accounting to 60 percent of the warming caused by anthropogenic emission as of 2000 (NAS, 2010). Indeed, water vapor is the most important greenhouse gas in terms of its proportion in the air. However, it has lower radiative properties and it stays for few days in the atmosphere as compared to the life cycle of carbon dioxide that might go up to centuries (Neu, 2009;

---

[9] Urban Heat Island effect is the relative warmth of big cities as compared to the surrounding rural areas. Big cities are most inhabited and have higher heat output, are built in less greenery areas and buildings absorb a large amount of energy and thus they exhibit higher temperature than the surrounding areas (Neu, 2009; Pittock, 2009; NAS, 2010; Lomborg, 2006).

[10] Poverty, AIDS, nuclear weapons, etc (Giddens, 2008; Lomborg, 2006).

NAS, 2010). Given such shorter life time and low radiation absorption qualities of water vapor, they maintain that carbon dioxide is more problematic than water vapor.

Physical scientists also claim an existence of a strong correlation between atmospheric concentration of greenhouse gases (carbon dioxide) and global warming (IPCC, 2007a; Tuckett, 2009; Garnaut, 2011). Earth's surface temperature has clearly increased over the past 100 years along with sharp rise in the concentration of human induced carbon dioxide and other green house gases (NAS, 2010). According to these scientists, the reason for such rise in the concentration of these gases is unquestionably human activity and there is no other plausible justification other than this (Helm, 2005). The American National Association of Sciences further supports this assertion through rigorous statistical procedures of observed climate trends (supplemented by analysis of climate models) which fail to link the warming with natural variations, particularly since the late 1970s.

IPCC (2007a) also confirms that observed sea level rise, and decrease in snow and ice extent are all consistent with global warming. Global average sea level rose at an average rate of 1.8 mm per year over 1961 to 2003, and this figure has enlarged to 3.1 mm per year in between 1993 and 2003 (IPCC, 2007a:30). Global average surface temperature has increased by around 0.7°C since the start of the industrial era and it continued to increase by 0.74±0.18°C in the 100 years between 1906 and 2005 (UNDP, 2007:6-7; IPCC, 2007:30a). Among the 12 years in between 1995 and 2006, 11 of the years rank amongst the top 12 warmest years since 1850. Global warming over the 50 years in between 1956 and 2005 has nearly doubled to that of the 100 years from 1906 to 2005 (IPCC, 2007:30a). Conversely, the total solar and volcanic activities during the past 50 years (up to 2007) would likely have created cooling effect (IPCC, 2007a). Therefore, it is concluded that most of the observed rise in global average temperature since the mid-20th century is very likely[11] due to anthropogenic concentration of green house gases (IPCC, 2007a; NAS, 2010; Garnaut, 2011). Further, IPCC (2007:45a) projects rises in global temperature in a range of 1.1 to 6.4°C and sea level in between 0.18 - 0.59 meters at the end of 21st century (2090 -2099) relative to 1980 - 1999 averages.

Skeptics again continue attacking some of the evidences following the 2009 hacking of emails (Climategate[12]) from Climate Research Unit (CRU) of the University of East Angelia, which also put into question some of the evidences published by the Intergovernmental Panel on Climate Change (IPCC). The emails were used by skeptics and critics to suggest that some scientists had used the data selectively to substantiate

[11] The likelihood reference, 'very likely', according to IPCC, refers to 90 - 99 percent of chance.

[12] Climategate is the name tagged by the media referring to leaked emails of 2009 from the Climate Research Unit of the University of East Angelia, United Kingdom (Garnaut, 2011). Climate Research Unit of the University of East Angelia is one of the leading scientific centers of climate change and is well known proponent of global warming as a human induced event.

the idea of global warming (Garnaut, 2011). Out of this concern, the Berkeley Earth Project[13] which gathered a team of 10 scientists[14] was established by Richard A. Muller, professor of physics at the University of California, Berkeley. Finally, they came up with results consistent with previously reported by scientific teams in the United States and Britain[15] (Rohde et al., 2011). They also concluded that urban heating effect (UHI) has nearly negligible effect on global warming (Wickham et al., 2011).

Though skeptics are dwindling in number and the great majority of scientists believe in the reality of climate change (Giddens, 2008), the debate might not be easily resolved in the coming decades due to uncertainties surrounding the science and the infancy stage of the science of climate change. The controversies will continue until scientific knowledge prevails with theories that are fully supported by scientific evidence (Zannetti, 1998). Even in the presence of scientific evidence, the debate might not be abated easily. This is because of the fact that the debate is not only a scientific wrangling between skeptics and believers, but also it is a phenomenon which sustains resentfully within believers and policy makers due to disagreements on the scale of and commitments to mitigation, and the amount and mechanism of financing adaptation, among other things (Schipper, 2007; McGray et al, 2007; Klein & Person, 2008; Harmeling & Bals, 2008).

Given such controversies, where would the heated debate take us or what implication does it reflect to local farming communities and the discourse itself? First, local communities will continue to suffer from the impacts while the discourse is supremely engaged in the debate and actions of response are neither agreed upon. Second, the discourse may lose the opportunity to grab local knowledge to supplement and validate some of its epistemological claims through concrete examples from localities while totally engaging with science knowledge at global level. Third, top-down policies and decisions that pay little attention to local knowledge and values may evolve (are also being imposed) and implementation might be marred by local community's rejection or to the worst might end up in destabilization of local response systems that enabled the co-existence of people with the environment.

---

[13] The Berkeley Earth Project is a scientific project established by Richard A. Muller, professor of physics at the University of California, Berkeley, following the 2009 hacking of emails from Climate Research Unit (CRU) of the University of East Angelia. The project was launched with the aim of resolving the criticism that followed the email hacking and prepare an open responsive source of records that will entertain further criticisms or points of discussion. The project was conducted using 39, 000 weather stations, which is more than five times that many climate studies have previously used (http://berkeleyearth.org/study/, accessed 11/01/2012).

[14] Source: http://www.bbc.co.uk/news/science-environment-15373071, accessed 10/01/2012.

[15] Source:http://berkeleyearth.org/pdf/berkeley-earth-summary-20-october-2011.pdf, accessed 11/01/2012.

Indeed, genuine debate is one of the healthy scientific inquiries to examine assumptions critically before reaching into conclusions. However, the existing debate is not only of due to the limits of the science, but also it is partly driven by economic and political interests (Pittock, 2009). Such embedded economic and political interests on the debates further divide the already divided world on responding to climate change. In such wave of debates and controversies, where actions of response are neither agreed, apparently it is local communities (such as farming societies who are dependent on natural resources) who will be most likely set aside and continue to suffer.

As could be figured out from the discussion so far, the science of climate change and policy debates entirely oscillate at the top level. On the contrary, the impacts of climate change are mainly observed at the grass-roots level, where the impacts might be meaningfully understood through the lens of local knowledge. Despite the importance of local knowledge in solving environmental and development problems (World Bank, 1998; Nyong et al, 2007), and exploit it as a potential source of knowledge to complement the science in the face of the divided world, neither does the discourse formally engage local knowledge into scientific and policy discussions. This is mainly the reflection of the following reasons.

In the first place, climate change discourse mainly focuses on global climate and lack of consensus on the causes (of climate change) apprehended the discussion to stay at higher levels. Secondly, climate models generally remain very weak at providing information at local levels while depicting bigger pictures of climate change at global and regional levels and its likely consequences in light of different scenarios[16] of human development (Salick & Byg, 2007). Third, despite a general consensus on mitigation and adaptation as the two mechanisms to combat climate change and variability, there is a difference on the scale and approaches of the responses, the amount and mechanisms of financing and operationalizing the principle of common but differentiated responsibilities among nations (Schipper, 2007; McGray et al, 2007; Klein & Person, 2008; Harmeling & Bals, 2008; Leggett & Lattanzio, 2009; Gibson & Loh, 2011), and hence discussions are largely engrossed around these issues. Fourth, traditionally perceived inferiority of local knowledge as compared to scientific mode of inquiry, although this perception has been dwindling in recent scientific publications with the acknowledgement of indigenous peoples' observations of environmental changes (Green & Raygorodetsky, 2010), has hampered the role of local knowledge in the discourse.

---

[16] Scenarios, known as SRES (Special Report on Emissions Scenarios), refer to plausible range of greenhouse gases emissions used as a basis for climate projections with underlying assumptions of various development pathways, which cover demographic, economic and technological changes and resulting greenhouse gases emissions (IPCC, 2007; Pittock, 2009).

As such, climate change discourse is based purely on scientific inquiry. Such approach, according to Jasanoff (2010), detaches knowledge from meaning, given that meanings emerge from embedded experiences and interactions with the social and natural environment (Jasanoff, 2010; Cobb, 2011). Jasanoff (2010:235) argues that "science is not the only, not even the primary, medium through which people experience climate". Senses supplemented by experience play an important role in understanding climate, and epistemic claims of climate change are most trusted when they engage practices and local values (Jasanoff, 2010). Therefore, it would be benefiting for climate change discourse, if it uses multiple sources of knowledge systems (epistemologies) that are composed of social and environmental interactions (Jasanoff, 2010; Cobb, 2011). In this regard, local knowledge, as a knowledge body founded on experience, and interactions with social and natural environment, could play a pivotal role in enriching the discourse and complementing scientific evidences and vice versa. Even if engaging local knowledge into climate change discourse is not an easy task, policy makers need to work with due consideration to local knowledge as it is vital in designing policies that are responsive to local climate impacts.

Local societies in the developing world, particularly rural farming communities, are highly detached from the scientific understanding of climate change. They either remain largely unaware of the global discourse of climate change. However, they have their own ways of climate detection[17] and attribution[18] which they acquired based on experiential learning and profound knowledge passed to them through generations. It is through their own mechanisms of climate detection and attribution that farming communities understand the climate, comprehend the impacts and guide their adaptive behavior. Therefore, understanding local knowledge of climate change detection and attribution (at the localities) certainly forms the basis in responding to climatic concerns at local levels. It might also help to enhance scientific inquiry in places where there is lack of scientific data (Green & Raygorodetsky, 2010; Kirkland, 2012) and enrich the general scientific findings by locally detailed context factors. It also provides tools for managing climate change (Kirkland, 2012) which might be of crucial importance in coping local vulnerabilities. From this view point, it plays an important role in designing policies that are responsive to local needs by reversing or balancing the top-down approach of decision making to deal with climate concerns at local levels. In terms of policy, it is this level that needs adequate focus.

---

[17] Climate change detection is the process of demonstrating that climate has changed in some defined statistical sense, without providing a reason for that change (IPCC, 2007a).

[18] Climate change attribution is the process of determining the most likely reasons for the detected changes (IPCC, 2007a).

2.2. Economic Approaches to Measure the Impact of Climate on Agriculture

Climate is a prime natural agent which plays the most important role in shaping agricultural production. Climate variability and change may affect agricultural production through changes in climatic conditions, mainly in temperature, precipitation and extreme events. These changes in turn alter soil moisture and fertility and the length of growing season (McGuigan et al., 2002), and thus affecting agricultural production.

Even though no country has immunity to climate variability and change, some countries which are highly dependent on agriculture are more vulnerable and bear enormous adverse impacts. For many developing countries, agriculture provides employment for 65 percent of the labor force (65 percent in Sub-Saharan Africa and 60 percent in South Asia), and accounts 20 - 60 percent of the GDP (Ludwig et al., 2007:8; UNCTAD, 2010:1; Hoffman, 2011:2). Through value chains, it also supports industries and services which contribute for more than 30 percent of GDP, even in affluent nations (Hoffman, 2011:2). Given such critical role of agriculture in human welfare vis-à-vis its sensitivity to climate, there is high concern about the impact of climate variability and change on agricultural production (Adams et al., 1998). Because of this, there has been a growing interest among agronomists, climatologists, economists and social scientists to measure the economic impact of climate on agriculture.

A number of approaches have been used to measure the economic impact of climate on agriculture. A simple classification commonly present in the literature categorizes the approaches in to two as either structural or spatial analogue approaches (Schimmelpfennig et al., 1996; Adams et al., 1998; McCarl et al., 2001). These approaches could be used both at micro- and macro-levels. Structural approaches are mainly crop simulation models carried out based on experiments, and spatial analogue approaches are regression methods that estimate the impacts based on cross-section data. To bridge the gap that could not be addressed by these two approaches, researchers have proposed other approaches based on neoclassical theories. The approaches are described as follows with due emphasis to spatial analogue approach, as this approach applies observed data collected from actual farm setting which is the interest of this study as well.

2.2.1. Structural Approaches

The structural approach comprises agronomic-economic or crop response simulation models which seek to find out the yield response of a specific crop under a combination of different climatic conditions and the results from simulations are fed into economic models of farmer behavior to determine production and prices (Dinar & Beach, 1998;

Seo et al., 2005). Although the approach brings together several disciplines such as atmospheric science, crop science and economics (Schimmelpfennig et al., 1996; Adams et al., 1998; Molua & Lambi, 2006), its foundation is largely rooted on natural science disciplines.

The central tenet of the approach stems from carefully calibrated physiological crop simulation models, which are framed on detailed laboratory experiments to construct the response of specific crops and crop varieties to different climatic conditions (Mendelsohn et al., 1994; Schimmelpfennig et al., 1996; Adams et al., 1998). Experiments are usually performed at micro-levels and could be allowed to integrate farm level adaptation such as changing planting dates and crop types and varieties, and adding or enhancing irrigation (Schimmelpfennig et al., 1996; Adams et al., 1998). Farming methods are kept similar across experimental conditions so that all the differences in outcome can be assigned to climate variables under consideration (Mendelsohn & Dinar, 1999). Once crop responses to changes in input mixes and climatic conditions are measured, yield estimates are then incorporated into economic models which seek either to minimize costs or maximize consumer and producer welfare subject to climatic and other constraints imposed on the model (Adams et al., 1998; McCarl et al., 2001). The economic variables might include land size, input supply and market prices.

This approach principally assumes that simulated climate effects are accurate and adaptation strategies of producers and consumers would proceed as modeled (McCarl et al., 2001). One of the limitations of this approach arises from this assumption. Indeed, the approach has advantages in the sense that it creates favorable conditions for detailed understanding of the physical, biological and economic adjustments that could follow changes in climate and other conditions assumed in the model (Schimmelpfennig et al., 1996). However, in the real world where climate could change abruptly and could also emerge with overwhelming extreme events that could in turn wither up agricultural production, accuracy assumption of climate effects simulated in the model will be in big challenge.

Profit maximizing (commercial) farmers might carry out compensatory or adjustment responses to changes in climate, such as planting drought resistant crops and increasing irrigation water when confronted with a decline in precipitation (Deschenes & Greenstone, 2006). Likewise, various socioeconomic factors and technological changes might dictate human behavior somehow. Consequently, producers and consumers, as rational beings, might more importantly adjust to these changes than the adjustments assumed in the model. In connection to this, several studies state that farmer responses considered in the approach are static or ad hoc and hence a range of modified adaptation strategies that farmers regularly make in response to changing economic and environmental conditions are ignored (Mendelsohn et al., 1994; Mendelsohn & Dinar, 1999; Kurukulasuriya & Rosenthal, 2003). For this reason, the models might tend to overestimate economic damages of climate change and underestimate adaptive capacity of farmers (García-Flecha & Viladrich-Grau, 2005).

Another drawback of the model is that, for aggregate studies, conclusions are extended to wider areas and diverse production systems based on relatively few experimental or simulation sites (Mendelsohn et al., 1994). Furthermore, as experimentation is extensive and expensive, simulation models are limited only to grain or selected crops and few geographic locations around the world (Mendelsohn & Dinar, 1999; Kurukulasuriya & Rosenthal, 2003; Seo et al., 2005).

While providing a useful baseline for estimating the impacts of climatic change on agricultural production (Mendelsohn et al., 1994), experiment based structural models are more important for scientific testing than applicability in local context especially in developing countries where experiments are very limited and economic models are poorly calibrated (Seo et al, 2005).

2.2.2. Spatial Analogue Approaches

While structural approach is predominantly experimental, spatial analogue approach, on the other hand, employs cross-section data to pursue statistical or econometric estimations that attempt to measure the impacts based on observed differences in land values, agricultural production and climate (Schimmelpfennig et al., 1996; Adams et al., 1998; McCarl et al., 2001). Principally, spatial analogue approach was developed to draw inferences about how commercial farmers in cooler regions could adopt agricultural practices of warmer regions if climate is warmed (Schimmelpfennig et al., 1996; Adams et al., 1998). As described by Adams et al (1998), the foundation of this approach lays on the assumption that farmers (in cooler regions) will be able and willing to adopt agricultural practices of farmers in warmer regions. Among other things, the approach takes into account other factors that affect agricultural production which are typically ignored in structural approach, like land quality (soil type or quality). Even though this helps to isolate the effects of climate change, it depends on the representativeness of data and the ability of statistical analysis to isolate confounding effects (Schimmelpfennig et al., 1996).

The single most widely applied model in spatial analogue approach is the Ricardian approach, also called cross-sectional or hedonic model. Since this approach has been quite extensively applied and dominant in various research undertakings and will be used as an input in deciding the empirical model of this research, it is discussed at some length below.

Ricardian Approach

The Ricardian model was originally conceived by Ricardo in 1817 who theorized land rent (value of land) as the reflection of net productivity of farmland under perfect competition (Mendelsohn & Ariel, 2003; Mendelsohn & Reinsborough, 2007). Based on this theory, Mendelsohn et al (1994) developed a model to measure the impacts of climate shifts on land values in the USA. Later it was applied in Latin American, Asian and African countries (Mendelsohn et al., 1994; Seo et al., 2005; Gbetibouo & Hassan, 2005; Ouedraogo et al., 2006; Mano & Nhemachena, 2006; Kurukulasuriya & Mendelsohn, 2007; Molua & Lambi, 2006; Deressa, 2007; Madison et al., 2007; Féres et al., 2008; Seo & Mendelsohn, 2008a, 2008b; Thapa & Joshi, 2010).

According to Mendelsohn et al (1994), structural approaches do not take account of the varieties of economic substitutions and adjustments that farmers perform to adapt to changing conditions in the climate. As pointed out earlier, such exclusion tends to overestimate the effects of climate change on production. Accounting this and other drawbacks, they developed the Ricardian model with due consideration to socioeconomic substitution and adaptive activities.

The Ricardian model assumes that under competitive markets, land value, measured by the present value of expected net revenues, is the reflection of efficient use of land (Adams et al., 1998). Efficient land use, in turn, is affected by climate conditions, among other factors. Hence, changes in land values reflect the value of climate change to agriculture if prices of output/land and inputs remain constant (Deschenes & Greenstone, 2006). When the climate changes or varies across geography, farmers also undertake adjustment activities that maximize farmland returns (they will adapt to varying climatic conditions that could have potential effects on returns) (García-Flecha & Viladrich-Grau, 2005). In other words, it presupposes the presence of crop and farmer responses to climate in observed cross-sectional data such that the biophysical and economic adjustments imposed by climate change have been made across varying agro climatic zones (McCarl et al., 2001). With this assumption, the model links the effects of climate on land values along with relevant socioeconomic variables and potential adaptation measures undertaken by farmers which are considered implicitly (Seo et al., 2005). Thus, the Ricardian approach regresses climate, socioeconomic and relevant environmental variables against the market value of farmland to obtain estimated coefficients. The coefficients reveal the economic value of each factor, including climate variables. These coefficients can then be used to predict agricultural impact of future changes in climate (Mendelsohn & Reinsborough, 2007).

The model is expressed by quadratic terms of climate variables:

$$Y = a + bT + cT^2 + + dP + eP^2 + fZ + \varepsilon_n \tag{2.1}$$

Where; Y represents land value or net farm income, T and P represent temperature and precipitation variables, Z represents a set of socioeconomic variables (capital, soil quality and other household characteristics), ε is an error term, and a, b, c, d, e and f are parameters. Although it is difficult to justify on theoretical grounds, the model assumes quadratic terms of precipitation and temperature (Fezzi & Bateman, 2012). The quadratic terms are meant to reflect the nonlinearity of the response function between land value or net income and climate variables; where the function will have either a U-shape (when the quadratic term is negative) or an inverted U-shape (when the quadratic term is positive) (Mendelsohn & Reinsborough, 2007).

Several studies have employed this approach both in developed and developing countries. Some of them analyzed the impacts of climate change retrospectively, and others focused on projections. The studies used either land value or net revenue of production as dependent variable to regress on a set of independent variables including climate. The following brief summary of the studies depicts the results of these studies.

Ouedraogo et al (2006) regressed net revenue of crops in Burkina Faso on climate variables (temperature and precipitation), hydrologic variable (runoff) and socioeconomic variables. They found out that crop net revenue is more sensitive to precipitation than temperature changes. Their study also shows that since the climate in Burkina Faso is already hot and dry, the scenarios of decreasing precipitation and rising temperature will be harmful to crop production. On the other hand, the scenarios of increasing both precipitation and temperature will be tolerable because positive effects of precipitation will compensate the damages incurred due to increment in temperature.

A study conducted by Mano & Nhemachena (2006) on smallholder farmers in Zimbabwe shows that net farm revenues are affected negatively by increases in temperature and positively by increases in precipitation. They also reported that dryland farms will be the ones that will be more affected as compared to irrigated farms.

A study by Gbetibouo & Hassan (2005) in seven field crops in South Africa indicates that production of field crops is sensitive to marginal changes in temperature as compared to changes in precipitation. Their result shows that while rise in temperature affects net revenue positively, reduction in rainfall is negative to net revenue.

Deressa (2007) regressed net revenue on climate, household, and soil variables in Ethiopian farmers. He reported that climate variables have significant impact on farmers' net revenue. Although the impacts differ in seasons, increasing temperature and decreasing precipitation in general have damaging effect for the Ethiopian agriculture.

Kurukulasuriya & Mendelsohn's (2007) analysis of the impact of climate change on cropland based on 11 African countries[19], where they regressed net revenue on water flow, climate, soil and economic variables, shows a fall on net revenue as precipitation declines and temperature rises in general. They stated that the impacts are not likely to be uniform across the continent and this depends on the existing features of the climate and the climate scenarios considered.

Using the same data employed by Kurukulasuriya & Mendelsohn (2007), Madison et al (2007) confirmed that African agriculture is vulnerable to climate change. They indicated that those countries with hot climate would suffer the greatest loss in productivity as compared to countries with cooler climate. Assuming perfect adaptation, they portrayed that by 2050 warmer countries will lose a considerable amount of their production. For instance, Burkina Faso and Niger will entail a loss of 19.9 percent and 30.5 percent production, respectively. On the other hand, relatively cooler countries such as Ethiopia and South Africa will suffer a production loss of 1.3 percent and 3 percent, respectively.

Thapa & Joshi (2010) analyzed the relationship between net farm revenue and climate variables in Nepalese agriculture. They run two regression functions: the first considering climate variables alone and the other including socioeconomic variables. Their findings show relatively low precipitation and high temperature impacting net farm income positively during the fall and spring seasons which are the major harvesting seasons in Nepal. But, in the summer, low temperature and higher precipitation have an increasing effect on revenue. This shows that the same climate variables could have differing impacts on net revenue in different seasons.

Seo et al (2005) examined climate change impacts on the net revenue of four most important crops in Sri Lankan agriculture. Their report portrays that the impacts of rainfall increases are predicted to be beneficial to the country as a whole in all five climate scenarios, but temperature increases are predicted to be harmful. Nationally, the impacts vary from - 20 percent to +72 percent depending on the climate scenarios. This shows that estimating impacts largely depends on climate scenarios considered.

Seo & Mendelsohn (2008b) estimated the impacts of climate change on the value of Latin American farm land using some scenarios and they demonstrated that farmers will lose their average income by 14 percent in 2020, 20 percent in 2060 and 53 percent

[19] The countries include Burkina Faso, Cameroon, Egypt, Ethiopia, Ghana, Kenya, Niger, Senegal, South Africa, Zambia, and Zimbabwe.

in 2100[20]. In addition, they reported that both large commercial farms and small household farms are highly vulnerable to climate change; with small farms more vulnerable to warming and large farms more vulnerable to shifts (increases) in rainfall. The results appear to be highly worrisome for Latin American Agriculture sector. But, they also advised readers to be cautious of the results since the results are susceptible to omitted variables and the analysis did not account technological and price changes, and carbon fertilization[21], which is the typical trait of Ricardian approach.

As illustrated by the review of the above studies, the Ricardian approach seems to be a popular model of application and well regarded for its ability of incorporating farmers' economic substitution and adjustments. The approach is also sound for the inclusion of different climatic zones to capture spatial adaptation as an initial response to climate change (Adams et al., 1998). Its utilization of actual agricultural and socioeconomic data and cost effectiveness as compared to structural (experimental) approach is additional advantage of the approach. However, quite a number of concerns pose serious questions on the approach. The following points pay attention to these limitations.

1. Lack of careful controlling mechanism to account variations across farms in various agro ecological zones and the state of covarying factors with climate. Unlike experimental approach, which relies on carefully controlled experiments, the Ricardian approach employs observed data from various farms in different agro climatic zones, and it is this agro climatic difference that builds the core of the model. As stated above, the model attempts to link variations in net land value or farm income with climate heterogeneity (variations in precipitation and temperature) inherited from differing agro ecological zones (Mendelsohn et al., 1994). However, farms might vary due to a number of factors in addition to climate, such as access to market, irrigation and soil quality. In fact, statistically controlling the factors might lessen the problem and some recent studies are taking this concern into consideration by including pertinent factors that might be sources of such concern (Deschenes & Greenstone, 2006). But, due to data limitation and measurement problems, estimates might be biased (Mendelsohn & Dinar, 1999). Even in the event of data availability for control variables, these variables are found to covary with climate variables. Deschenes & Greenstone (2006) in their study of the impacts of climate change on agricultural land in the USA found out that estimates from Ricardian approach are unreliable because of the covariance of control variables (soil characteristic, population density, per capita income and latitude) with climate variables (temperature and precipitation). So, the approach is criticized for its inability to fully control the effect of other factors that could also be responsible for variations in land value or net farm income (Mano & Nhemachena, 2006; Thapa & Joshi, 2010).

---

[20] Estimated under the severe Canadian Climate Center (CCC) scenario.

[21] Carbon fertilization is predicted to increase future productivity, especially in temperate zones.

2. It fails to consider price changes. Mendelsohn & Dinar (1999) acknowledge that the model does not provide means to capture price changes. But they argue that since prices in most cases are determined by global market and some studies[22] predicted small effects of climate change on aggregate (global) supply, the effect from the assumption of constant price is minimal. However, according to Schimmelpfennig et al (1996), this assumption is potentially a serious drawback of the Ricardian approach. They stressed that climate change could cause large and widespread changes in prices. Prices could also change from changes in local supplies and global markets. Therefore, changes in price would lead to incorrect estimations of land values, and hence might plague the whole process of the approach. In connection to this, Seo et al (2008) pointed out that if local prices are to change due to local circumstances, estimates from the approach will be biased.

In a different perspective but related to price changes, farmers might also adjust farm level decisions based on price changes. However, as there is no mechanism to control such price induced changes, farmers' responses are generally regarded as climate change induced.

3. It presumes farmers as economic agents who are able to detect changes in climate so that they respond automatically (Mendelsohn et al., 1994; Adams et al., 1998; Mendelsohn & Dinar, 1999). One of the major problems in measuring agricultural impacts of climate change is lack of methods to capture farm-level adaptation undertaken by farmers (Darwin, 1999). Offsetting this problem is one of the strong points of Ricardian approach. However, the Ricardian approach captures only the effect of adaptation implicitly through the use of net revenue by assuming that farmers will automatically respond to changes in climate. Adaptation is dependent on condition of changes in climate, detection of these changes, risk assessment by farmers and as well as enabling socioeconomic and institutional conditions. Hence, adaptation might not occur even if changes are happening in the climate. Detailed explanation in this topic will be provided later in section 2.3. For the time being, it suffices to show that adaptation might not occur due to changes in climate per se. Several studies also show that after detecting changes in local climate, quite a lot of farmers fail to undertake adaptive strategies due to several reasons (Maddison, 2006; Deressa et al., 2009; Gbetibou, 2009).

4. It assumes that adaptation is a costless activity. Alike structural approaches, Ricardian approach also assumes that adaptation is a costless process (Adams et al., 1998; McCarl et al., 2001; Molua & Lambi, 2006). In reality, farm level adaptation itself is associated with several cost incurring resources such as fertilizer, improved seeds, irrigation and other socioeconomic resources.

---

[22] They cited Reilly, Hohmann & Kane (1994) and Reilly et al (1996).

5. It fails to keep its central assumptions in the context of developing countries. The Ricardian approach works if and only if functioning markets are in place. It is with this basic assumption that it was first applied to measure the impacts of climate change in the USA, where well- functioning markets exist.

Under competitive markets, it is possible to get the market value of farmland. Land as a market good also warrants the existence of private land ownership. However, in developing countries, markets are inefficient and hence it is difficult to obtain land values. Not only is market a hindrance to valuation of land, but also land tenure itself is a major obstacle in countries like Ethiopia where land is owned by the government and thus market value of land cannot be reflected at all. In fact, to address this problem, net farm revenue is used as a proxy to land value, in which case the Ricardian approach losses its flavor and assimilates with the standard neoclassical production function. Still substitution of land values with net farm revenues or income demands the assumption of competitive market which stimulates farmers to adjust to markets so as to maximize net incomes (profit) from production. But, the majority of farming households in developing countries are subsistence, i.e., they are largely the sole consumers of own production. Apparently, subsistence farmers operate largely neglecting existing market conditions (Kopeva & Noev, 2003). Consequently, the market assumption remains weak among subsistence farmers as far as production is channelized largely to the fulfillment of the households' consumption needs. Due to this, Ricardian approach fails to establish the link between market assumption and consumption oriented subsistence farming.

On top of this, the approach excludes household labor in computing net revenue since household members work on family farms with no wages. Although labor is widely acknowledged surplus in the agricultural sector of developing countries, economic theory still maintains labor at the heart of the production function. Nevertheless, in applying the Ricardian model among African farmers, Kurukulasuriya & Mendelsohn (2007) and Mano & Nhemachena (2006) excluded household labor acknowledging that the inclusion would result in negative net revenue. This noticeably strengthens the consumption orientation of subsistence farming. Otherwise, with the assumption of net revenue as a profit, there cannot be agricultural practice that persistently operates under loss. That is why Schimmelpfennig et al (1996) explicitly stated in their explanation about spatial analogue approach (Ricardian approach) by using the word 'commercial farmers' to assert the behavior of farmers that are governed by competitive markets such that the assumption of Ricardian approach would be met. Therefore, the application of Ricardian approach in subsistence farming is problematic, in particular, and less applicable in most developing countries, in general, as it requires strong assumptions which are ill-represented or not available in these countries.

That is why Maddison et al (2007) expressed for the acknowledgment of the weaknesses of the Ricardian approach in estimating impacts. Similarly, Seo & Mendelsohn (2008b) expressed their concern while interpreting the results of Ricardian approach due to its weaknesses (which includes susceptible to omitted variables, and disregarding technological and price changes, among other things). Therefore, future research has to endeavor to construct models that have sound theoretical background and are statistically robust (Deschenes & Michael, 2006) and appropriate in the context of developing countries, as these nations are the most affected by climate change and variability, and are less studied ones.

2.2.3. The Production Function

Production functions of various forms are customarily used to specify agricultural output to the combination of inputs. The most popular form of agricultural function (Cobb-Douglas form) can be expressed as:

$$Y = AL^{\alpha}Ln^{\beta}K^{\gamma} \quad (2.2)$$

Where Y is output and, L, Ln and K are labor, land and capital inputs respectively. A denotes total factor of productivity and α, β and γ are the corresponding parameters to labor, land and capital.

By augmenting or directly adding climate variables as inputs into this model (in such form of $Y = \mathrm{AL}^{\alpha}\mathrm{Ln}^{\beta}\mathrm{K}^{\gamma}\mathrm{W}^{\delta}$, where W represents climate, usually temperature and precipitation), researchers attempt to quantify the impact of climate on agricultural production. One can find very easily a vast application of such model in the literature. Despite its extensive application, such approach entertains very basic weakness. It lacks theoretical basis to equate climate as an input that directly affects output. As a matter of fact, climate influences production indirectly via factors of production, i.e., by its influence on soil (land) moisture and temperature, and the physiological growth of the crop plant (which is very difficult to capture through the production function as a direct input). Such that climate variables such as precipitation and temperature allow agricultural output to increase or decrease without any additional use of factors of production (land, labor and capital) by influencing the productivity level of these factors. Such behavior of climate will thus affect agricultural output essentially through total factor productivity by altering the productivity of the production factors. Therefore, considering climate as an input along with conventional inputs cannot rightly show its authentic influence over agricultural output.

Another application of the production function considers the effects of climate on yield variability. With the exception of few experimental studies (Mearns, et al., 1997; Chen et al., 2004), all the approaches discussed so far (production function, structural and spatial analogue approaches) exclusively focus on the response of mean agricultural production to the shifts in mean values of climate variables. Such considerable attention on average agricultural production has provided only limited information on crop yield variations in response to climate change and variability (Mearns, et al., 1997; Chen et al.,

2004; Isik & Devadoss, 2006). Since agriculture is sensitive to changes in climate factors, variability in production is also an important aspect of agriculture. To address this concern, researchers have been using a model commonly known as Just and Pope's production function, which extends the production function specified in equation 2.2 above.

Just and Pope's production function was developed in the late 70s to determine production risks econometrically (Just and Pope, 1979). Risk considerations are essential in the assessment of crop production (Ligeon et al., 2008). To account risk, Just and Pope proposed a function that helps to identify risk increasing and risk decreasing inputs (Ibid). The model allows inputs that might have positive effects on mean production to impact positively or negatively the variance of yields, and hence helps to interpret whether a particular input is risk increasing or decreasing (Just & Pope, 1979; Di Falco, et al., 2006; Ligeon et al., 2008).

The basic idea behind Just and Pope's production function revolves around constructing the production function as a component of two functions. The first component of the function is related to the mean output, and the second one is related to the variability of the output (Just & Pope, 1979; Koundouri & Nauges, 2005).

As per the specification presented in Saha et al (1997), Just and Pope's production function could be expressed as:

$$\boldsymbol{Y} = \boldsymbol{f}\,(\boldsymbol{X};\ \boldsymbol{\beta}) + \boldsymbol{u}\,; \qquad (2.3)$$

Where; $\boldsymbol{u} = \boldsymbol{h}\,(\boldsymbol{Z};\ \boldsymbol{\alpha})\boldsymbol{\varepsilon}$; and Y is the average or the mean output, X and Z are vectors of explanatory variables which may have commonalities, u is heteroskedastic disturbance term, β *and* α are the corresponding parameters. The first functional equation, $f\,(X;\ \beta)$, represents the mean or average function (deterministic component) where mean yield is explained by variables given by X. The second part of the equation, u which is specified as $h\,(Z;\ \alpha)\boldsymbol{\varepsilon}$, is the variance function (risk component) which captures the effect of inputs Z on the variance of the output, and ε is a random error distributed with zero mean and variance σ^2. These two equations allow inputs to affect both mean yield ($\boldsymbol{E(y)} = \boldsymbol{f}\,(\boldsymbol{X};\ \boldsymbol{\beta})$); and yield variance ($\boldsymbol{Var(y)} = \boldsymbol{\sigma_i}^2\boldsymbol{h}\,(\boldsymbol{Z};\ \boldsymbol{\alpha})$ independently.

The model is constructed based on the assumption that the heteroskedasticity[23] nature of the variance is related to all or some of the explanatory variables. Thus, the variance function, which is supposed to capture production risk, is derived from the presence of heteroskedasticity (Just and Pope, 1979). Since the production function does not presume any a priori restriction on the risk effects of inputs, an input may be classified as risk-increasing if $\partial h(Z;\ \alpha)/\partial Z_i > 0$ or risk-reducing if $\partial h(Z;\ \alpha)/\partial Z_i < 0$ or risk

[23] For the term's description see Section 5.3, Chapter 5.

neutral if $\partial h(X;\ \alpha)/\partial Z_i = 0$). Therefore, if the variability of the output decreases as the amount of input increases, such an input is considered as risk-decreasing, and when the variability of output increases as the amount of input increases, the input in this case will be regarded as risk-increasing.

Given the problem of heteroskedasticity, the model is estimated either by a three stage feasible generalized least square (FGLS) or the maximum likelihood (ML). Although FGLS tends to dominate the literature, some studies show that maximum likelihood method is more efficient than FGLS in small data setting (Saha et al., 1997).

Unlike the Ricardian approach, this model enables simultaneous estimation of the effects of climate variables and other factors on mean yield and the variation of yield. Using this advantage of the model, Chen et al (2004) in major agricultural crops in the U.S., Isik & Devadoss (2006) in some selected crops in the state of Idaho (USA) and Cabas et al (2010) in Canadian agriculture have applied the model in explaining the effects of climate variables both on the mean and the variance of agricultural output.

Chen et al (2004) indicated that changes in average climate conditions cause alterations in crop yield levels and variability in crop specific manner. They reported that increased rainfall enhances corn yield, while decreasing yield variance; and temperature having the reverse effect. In the case of Sorghum, more rainfall increases both yield and yield variance; and higher temperature reduces both yield and yield variability.

Isik & Devadoss (2006) found out that climate change will have modest effects on the mean crop yields, but will significantly reduce the variance and covariance for most of the crops considered in the study. They concluded that climate change will have differential impacts on mean crop yields, yield variability and covariance of crops so that famers will likely expand the acreage of crops whose mean yield increases and/or variability decreases in response to climate change.

Even though Just and Pope's production function seems appealing by enabling to capture the effect of climate variables and other inputs on mean yield and yield variance at the same time, it is criticized for its attempt to link variance with risk. In this regard, Rossi (1984) argued that an increase in variability of output does not correspond to increased risk[24], and increased usage of an input may not, necessarily, enlarge the variance of output.

[24] Rossi (1994) cites Rothschild & Stiglitz (1970), and indicated that increase in variance does not correspond to increased risk as per the conceptualization of Rothschild & Stiglitz (1970). For further discussion, see Rossi (1994).

2.3. Adaptation to Climate Variability and Change

The two universally accepted mechanisms to combat climate variability and change are mitigation and adaptation (Smit et al., 1999; Wall & Marzall, 2006). Mitigation is an action to reduce emission of green house gases (GHGs) while adaptation intends to moderate the adverse effects of climate variability and change through a range of specific actions (Füssel & Klein, 2002; Füssel, 2007). Borrowing medical terms, one can simplify mitigation as a direct curative treatment and adaptation as a prescription for the symptoms.

Policy formulation and research in the field has been carried out in these two lines of response mechanisms since their inception at the 1992's Rio Earth Summit (Bahinipati, 2009). Nevertheless, mitigation has traditionally received much attention both from scientific and policy viewpoint while pushing adaptation to play a marginal role in the discourse (Kates, 1997; Pielke, 1998; Klein, 2001; Füssel & Klein, 2002; Levina & Tirpak, 2006; Wall and Marzail, 2006). The reasons behind this could be classified into two.

The first is a theoretical reason that goes with the existence of two distinct schools of thought, which govern human response to climate change. Kates (1997) termed these schools as preventionist and adaptationist. While these schools oppose in their notion of the role of adaptation, they converge in the sense of discouraging adaptation. A preventionist school believes in the catastrophic scenery of climate change, and proposes drastic emission reduction measures. For them, adaptation plays a weakening role on society's willingness to reduce the emission of GHGs and thus retreating mitigation efforts. On the other extreme, adaptationist school argues that both natural and human systems have lived for centuries naturally adapting to changing environmental conditions, and hence no need for mitigation nor systematic (planned) adaptation. For this school of thought, planned adaptation interferes with the sovereign adaptation process of natural systems (Klein, 2001; Füssel & Klein, 2002).

The second is a technical reason that goes with the points identified by Füssel (2007). Mitigation reduces the root causes of climate change. Consequently, it has the ability to reduce impacts in all climate sensitive systems. On the other hand, adaptation is limited to specified or targeted systems and hence the efficacy of proactive adaptation remains uncertain due to uncertainties in climate change. In terms of monitoring, mitigation is comparatively easy to establish binding levels and assess them quantitatively than adaptation, which is blurred due to lack of effective measurement techniques.

In spite of these arguments for substantial emphasis to mitigation, other convincing propositions for adaptation emerged subsequently. Of course, adaptation is not new in the history of societies. It existed along with the existence of human beings and societies have adapted to climate variability and change over the course of human history and will continue to do so (Adger, 2003). However, in this era where climate variability and change are supposed to impose new stresses on both natural and socioeconomic systems (Berkhout et al., 2004), adaptation came to emerge in climate change lexicon as a systematic concept that should not be left to its customary way of evolution.

One reason for adaptation goes with incremental understanding of unavoidability of climate change due to the long life span of already emitted GHGs in the atmosphere (IPCC, 2007b). Since the earth is already exposed to past emissions and these emissions can no longer be prevented, adaptation is regarded as the only choice to offset risks associated with these past emissions. The other reason which twisted research and policy considerations towards adaptation is linked with the growing incidents of climate related disasters in various parts of the world. The outcome of mitigation by nature cannot be realized immediately; even in the presence of aggressive emission reduction strategies, the upshot of mitigation will not disembark to the pressing effects of current climate variability and change. Societies in many parts of the world are already under increased vulnerability from population growth and technological change. Climate impacts exacerbate this vulnerability (Pielke, 1998). In this case, adaptation plays a double role: while addressing risks associated with current climate problems, it also helps to boost the resilience of human and natural systems to future risks of climate change.

Finally, the birth of a third school of thought, known as realist, has also played an important role in advocating inquiries for adaptation. According to this school, climate change is fact and thus adaptation is a decisive and pragmatic response option alongside mitigation (Klein & MacIver, 1999). In its report of 1996, IPCC (1996, in Pielke, 1998) strengthened this view by articulating adaptation as a very powerful option to complement mitigation efforts. Consequently, the scientific community in the last decade has paid a good deal of attention to adaptation as a proactive and reactive response especially aligned to the developing countries which are said to be more vulnerable to climate change (Huq et al., 2003).

2.3.1. Conceptualizing Adaptation

In fact, adaptation is now regarded as one of the fundamental response strategies to combat climate change and variability. But, how is it conceptualized in the context of climate change and variability?

Scholarly exercise in the field of evolutionary biology pioneered the systematic study of adaptation in the early 19th century. In the early 1990s, anthropologists were attracted to the term with due interest to understand human behavioral adjustment to environmental changes. When it comes to climate change discourse, adaptation is a late phenomenon that emerged only in the 1990s (Janssen & Ostrom, 2006). Though, it is a latecomer to the discourse, one can find a number of definitions in the literature.

Pielke (1998:159) defines adaptation as "adjustments in individual, group, and institutional behavior in order to reduce society's vulnerabilities to climate". Similarly, Brooks (2003:8) defines it as "adjustments in a system's behavior and characteristics that enhance its ability to cope with external stress". These definitions are similar in referring adaptation as an adjustment in behavior. But they differ slightly in the sense what adaptation delivers. The first definition directly links adaptation with its ability to minimize society's susceptibility to external effects caused by a wider climate (variability and extreme events, not just only climate change). On the other hand, the second definition is more interested on capacity boosting characteristics of adaptation to external stress. Brooks's (2003) reference of 'external stress' implies the existence of 'internal stress', which denotes 'social vulnerability[25]' in his words. Climate variability and change exacerbate internal stress. A system, at first place, copes with external stress when it is able to reduce its internal (inherited) vulnerability. Hence, adaptation, according to Brooks (2003), is not merely an adjustment that focuses on reducing a system's vulnerability to external stress but also of social vulnerability (to increase adaptive capacity[26]).

Quite well cited definition of IPCC (2001:982) describes adaptation as "adjustment in natural or human systems in response to actual or expected climatic stimuli or their effects, which moderates harm or exploits beneficial opportunities". While sharing in its reference of adaptation as an 'adjustment' with the above two definitions, this definition explicitly mentions three specific things. First it stresses the existence of actual or anticipated climate conditions. Second, it implies the need for adjustment in natural and human systems to these climate conditions. Lastly, it refers adaptation not only meant to restrain harm but also a systemic apparatus to use opportunities that might emanate out of changes in the climate. For instance, climate change is expected to increase heat waves and agricultural production in temperate areas, while reducing production in tropical areas (IPCC, 2007a). On the one hand, adaptation serves to reduce production loss in tropical areas; on the other hand, it helps to benefit from production boost in temperate zones while necessitating adaptation to heat waves in this climate zone.

Another definition notes adaptation as "any adjustment, whether passive, reactive or anticipatory, that is proposed as a means for ameliorating the anticipated adverse consequences associated with climate change" (Stakhiv, 1993 in Smit et al., 1999). This definition basically focuses on listing various forms of adaptation targeted to improve undesirable conditions that might arise due to climate change.

The above definitions could be easily linked with the framework of adaptation outlined by Smit et al (1999). According to Smit et al (1999), systematic conceptualization of adaptation should comprise three basic questions. The first question addresses

---

[25] See chapter 4.

[26] See chapter 4.

'adaptation to what', which refers to the specific manifestations of climate variability and change. Pielke's (1998) reference of 'climate' and Brook's (2003) 'external stress', Stakhiv's (1993) of 'climate change' and IPCC's (2001) 'actual or expected climate stimuli' symbolize this question. The second question refers to 'who or what adapts', which exactly implies to 'individual, group and institutional behavior', 'systems', and 'natural and human systems' respectively in each of the three first definitions in order of their presentation above. Finally, the third question addresses the 'how of adaptation', which refers to the processes and forms of adaptation. This question is implicitly addressed rather by Stakhiv's (1993) terms of 'passive, reactive or anticipatory'.

Literally, 'adaptation to what' and 'who adapts' are quite visible to answer. However, a highly important question and at the same time a complex one is the 'how of adaptation', which involves forms and processes of adaptation. It is in this question that the factors behind adaptation, such as socioeconomic and institutional factors that motivate adaptive behavior, are dealt with.

Adaptation might take many forms. It could be autonomous[27] or planned[28] in terms of purpose; it might be classified as short term/tactical[29] or long term/strategic[30] in terms of temporal scope; it can also be either localized or widespread[31] in terms of spatial scope (Smit et al., 1999).

Whether adaptation is short term or long term, or localized or widespread, or autonomous or planned, in terms of timing it would be either proactive or reactive. Thus, any classification of adaptation comes under the umbrella of proactive or reactive typology. As can be easily inferred from their names, proactive adaptation is an adaptation that takes place to anticipatory climate stimuli; whereas reactive adaptation refers to an adaptation that takes place in response to already observed climate stimuli (IPCC, 2007b). This classification in terms of timing coincides also with adaptive behavior of natural and human systems. Adaptation within natural and ecological systems is reactive in nature, but it could be proactive or reactive when it comes to human systems (Jones, 2010). Adaptation within human systems is promoted either by

[27] Autonomous adaptation is the form of adaptation stimulated by ecological, market or welfare changes in human systems and which does not constitute a conscious response to climate stimuli (IPCC, 2007).

[28] Planned adaptation refers to a deliberate adaptation action by policy decision (IPCC, 2007).

[29] Tactical adaptation is an adjustment to immediate climate stimuli made in one season. Examples might include selling of livestock and purchasing feed (Smit & Skinner, 2002).

[30] Strategic adaptation refers to structural changes in the farm operation or management that would apply for a subsequent seasons or a longer term. Examples include changes in land use, crop type and use of insurance (Smit & Skinner, 2002).

[31] Localized adaptation is adaptation at local or national levels as compare to widespread adaptation at international level.

private decision makers (individuals, households and private companies) or public (government) interest (Smit & Pilifosova, 2001).

Table 2.1. Forms of Adaptation with examples

		Proactive	Reactive
Natural and Ecological Systems			• Changes in the length of growing season • Changes in ecosystem composition • Migration into wet lands
Human Systems	**Private**	• Purchase of insurance • Construction of house on stilts • Redesign oil-rigs	• Changes in farm practices • Changes in insurance premiums • Purchase of air-conditioning
	Public	• Early-warning system • New building codes and design standards • Incentive for relocation	• Compensatory payments and subsidies • Enforcement of building codes • Beach nourishment

Source: Klein (2001:16)

From the aforementioned discussion and information presented in Table 2.1., it is possible to infer that adaptation could be regarded as any proactive or reactive action undertaken by natural and human systems as a response to anticipated or observed climate stimuli. Its purpose is either to reduce vulnerability and harm by boosting adaptive capacity of natural and human systems and/or enable these systems to use the advantages that might surface from changes in the climate.

2.3.2. Adaptation to Climate Variability and Change in Agriculture

Historically, adaptation of agricultural systems to environmental changes has been the norm rather than exception and societies have historically adapted to climate conditions (Adger, 2003; Lotze-Campen & Schellnhuber, 2009; Baethgen et al., 2004). However, the current speed of climate change is inducing and modifying known variability patterns beyond the coping capacity of producers (FAO, 2008). The increasing frequency and magnitude of extreme weather events coupled with unprecedented changes in the climate is also imposing new and potentially overwhelming pressure on the capacity of existing adaptation strategies (Ziervogel et al., 2008). Therefore, deliberate and conscious adaptation that can cope with these evolving impacts is an immediate concern in agriculture.

Adaptation serves to buffer agricultural impacts from the modifications in the climate and will also help to improve the resilience of agricultural systems to uncertain climate impacts (FAO, 2008). The essentiality of agricultural adaptation, particularly in developing countries, is self evidenced by agriculture's multiple roles in these countries. Food security, employment, income and significant portion of GDP are drawn from agriculture. For instance, more than 60 percent of the people in Africa directly depend on agriculture for their livelihoods (FAO, 2003 in Ziervogel et al., 2008:20). The most serious problem of such overwhelming reliance on agriculture is the dependence on rain-fed agriculture which is by no means immune to climate stimuli. According to IPCC, unless effective adaptation strategies are carried out timely, some African countries could lose up to 50 percent of yield from rain-fed agriculture by the year 2020 (IPCC, 2007a:50). It further states that agricultural production and access to food will be severely compromised in many African countries. Such impacts that significantly undermine the prominent role of agriculture in food production and economic growth predominantly signify the criticality of adaptation.

Adaptation in agriculture usually takes place at two broad scales: macro- and micro-levels (Kandlikar & Risbey, 2000). Adaptation at macro-level deals with adjustments of agricultural production systems at national and regional levels vis-à-vis domestic institutions, international policies, climatic factors, markets and other strategic issues. Whereas, at micro-level it is concerned with adjustments and decision making at farmer level (Risbey et al., 1999; Kandlinkar & Risbey, 2000; Kurukulasuriya & Rosenthal, 2003; Nhemachena & Hassan, 2007). Since it is at micro-level that both the impacts of climate and adaptation are quite well observed (and adaptation eventually occurs), adaptation at farmers' level is detailed here under. In addition, as the interest of this study is at this level, paying attention to the level is worthwhile.

At farm level, climate variation, extreme events and shifts in climate can have considerable impacts on farm production, and hence food security. These climate stimuli drive farmers to adopt new practices or increase the intensity and quality of traditional adaptation strategies. Farmers may implement various types of adaptation strategies which might include changing planting dates, adopting a range of crop varieties/species and drought resistant crops, water harvesting, planting trees, increased usage of fertilizer, irrigation and conservation techniques, and some other farm practices which might vary in location and time.

Farmers who largely depend on natural environment are obviously fragile to climate variability and change, and on the contrary, they have low adaptive capacity. However, as rational economic agents, they have their own ways of adaptation mechanisms that involve processes that govern their adaptive behavior. An understanding of the processes (decision making) is fundamental as it helps to explore possibilities in dealing with climate variability and change and provide knowledge for better choices about how to attain more efficient and effective adaptation. Yet, the process of adaptation in agriculture remains to be rarely researched area (Smit & Skinner, 2002).

2.3.3. Theoretical Approaches to Agricultural Adaptation at Micro-Levels

In dealing with the process of adaptation to climate variability and change at farmer level, two important questions have now become the major interest areas that underpin adaptation research. These questions are: Under what conditions could adaptation takes place or what factors influence adaptive behavior? What is the impact of adaptation in the face of increased threat from climate variability and change? To understand the processes underneath these questions, researchers have followed some theoretical approaches and models so as to inform policy and infer important lessons to adaptation under projected climate change deep in the future.

Agronomic-economic and Ricardian approaches have been used mostly to measure the impact of adaptation. As regard to analyzing adaptive behavior, frameworks borrowed from agricultural technology adoption models are quite widely used. Theoretical frameworks developed from psychosocial theories are also used in very rare cases, although seem to be decisive.

2.3.3.1. Applications Based on Agricultural Technology Adoption Framework

Adoption of agricultural technology is an important model in agriculture which describes the process of adoption of high yielding technologies among farmers. It attempts to predict adaptive behavior (both the decision to adopt and adoption intensity) of farmers by assessing their personal and socioeconomic characteristics, the time factor and attributes of the technology (Padel, 2001; Smit & Skinner, 2002) using logit[32] or probit models. While adoption refers to farmer's use and non-use of technology at a period of given time, intensity of adoption denotes the extent to which the technology has been applied by the adopters (Charles, 2004).

Many studies have shown that socioeconomic characteristics of farmers with a number of institutional factors and attributes of a technology affect farmers' decision to adopt and adoption intensity (Padel, 2001; Charles, 2004; Onemolease & Alakpa, 2009). The studies designate attributes of technology to basic features such as compatibility, complexity and profitability of the technology (Charles, 2004); and socioeconomic and institutional factors to comprise age, gender, farming experience, labor availability,

[32] Refer chapter 6 for the description.

farm size, livestock ownership, education, access to extension service, access to credit, access to market, participation in social organizations, farm and non-farm income and tenure status (Charles, 2004; Rezvanfar et al, 2009; Oboh & Kushwaha, 2009).

Given the conceptual similarities between agricultural technology adoption and adaptation to climate variability and change (Deressa, 2010), several studies employed agricultural technology adoption models in the context of climate variability and change. Accordingly, adaptation studies regressed socioeconomic and institutional factors in the pursuit of investigating the factors that influence adaptive behavior of farmers to perceived climate variability and change.

These studies assume that adaptation is a two-step process where farmers perceive changes (detect signals) in the climate and then carry out adaptation strategies. For instance, Madisson (2006:6) presumed adaptation involving these two-stages: "first perceiving that climate change has occurred and then deciding whether or not to adopt a particular measure". Acknowledging the two-step process of adaptation, Gbetibouo (2009:1) similarly conducted a research on South African farmers with a hypothesis: "farmers adapt to perceived climate change and variability". Deressa (2010:89) also conducted his study with a premise that adaptation "initially requires farmer's perception that climate is changing and then responding to changes through adaptation".

It is quite customary to see the application of this approach in a number of studies (Maddison, 2006; Gbetibou, 2009; Hassen & Nhemachena, 2008; Deressa et al, 2009; Basu, 2011). These studies applied logit or probit models to identify the factors that determine adaptation in general, and discrete choice models (multinomial logit or probit model) when analyzing the determinants of farmers' choice of adaptation strategies. Accordingly, Madisson (2006) found out that while experienced farmers in Africa are more likely to perceive climate change, it is education, extension service and proximity to market that have positive impact on propensity of farmers' adaptation. He recommended particular incentives or assistance for those farmers who perceive climate change but fail to respond.

In his study in South Africa, Gbetibou (2009) indicated that although farmers' perception of climate change is in line with meteorological records of the weather stations, only 30 percent of the farmers who perceived the changes had applied adaptive strategies. His study highlights factors such as household size, farming experience, wealth, access to credit, property (land) rights, off-farm activities and access to extension quite important for adaptation.

Deressa (2010) showed the level of education, household size, gender, livestock wealth, extension service, access to credit and temperature as influencing farmers' propensity to adapt among Ethiopian farmers. He also indicated that lack of information on adaptation methods and finance are the main barriers to adaptation.

These studies obviously offer some insights about farmers' perception of climate change and possible factors that might influence adaptation. However, entire emphasis on socioeconomic and institutional factors neglects the role of psychological factors (risk perception) in guiding adaptive behavior. Social science research has demonstrated the strong influence of risk perception on people's decision in responding to hazards and environmental stress (Mileti 1993; Sánchez & Hernández, 1995; Kabat et al, 2002; Dwyer et al, 2004; Jenkin, 2006; Lupton, 2006; Plapp & Werner, 2006; Leiserowitz, 2007). Technology adoption studies in agriculture as well revealed perception of farmers to the attributes of technologies being influential in the decision to adopt (Adesina & Baidu-Forson, 1995). Logically, unless some level of risk perception is developed, simple perception of climate variability and change (signal detection) cannot necessarily lead to adaptation, even in the availability of enabling socioeconomic and institutional factors. For instance, in Gbetibouo's (2009) study, a large number of farmers perceived changes in the climate (precipitation and temperature), nevertheless only a bit larger than a quarter of them had taken adaptive measures while most of them were expected to have taken adaptation measures under the premise of the two-step process of adaptation.

In explaining as to why farmers fail to carry out adaptive actions, one may directly set out to the analysis of resource related factors, and may conclude that if farmers fail to promote adaptation while cognizant of climate variability and change, the reason behind this rely on socioeconomic and institutional constraints. The same is true with the above studies (Madisson, 2009; Gbetibouo, 2009; Deressa, 2010) which recommended socioeconomic incentives and assistance to enhance farmers' adaptive behavior. Such approach is partly driven by the influence of traditional analysis in the field which has laid its foundation on resource based assessment paying little or no consideration to psychological factors (Grothmann & Patt, 2005).

The importance of socioeconomic and institutional factors (onwards termed as SEI in this study) in facilitating and determining adaptation is beyond doubt. After all it is the presence and scarcity of such factors that draw a demarcation line between developing (characterized by weak adaptive capacities) and developed nations (characterized by strong adaptive capacity), the rich and the poor, and the vulnerable and the resilient. Nevertheless, when it comes to the response of individual behavior to risk-embedded stimuli, the existence of these factors per se might not necessarily lead to adaptive behavior. Even in some cases, implementing an adaptive strategy might not require a bundle of resources; for instance, changing planting dates based on the timing of rainfall.

It is the association of perception of climate variability and change with risks (risk perception) that is decisive in stimulating adaptive behavior rather than simple perception (signal detection) alone. In line with this, Protection Motivation Theory, which is discussed below, posits that individuals respond to risks if they perceive that they are vulnerable (susceptible) to risks; that they perceive the risks to pose potentially serious consequences; that they believe in the efficacy of counteractive actions (adaptive behavior); and that they believe in their capacity to implement the behavior. It is these cognitive processes that at first place motivate adaptive behavior and even before one thinks of resources.

The other drawback in this approach is linked with methodological problem in the inventory of adaptation strategies employed by farmers. The empirical studies already mentioned above regarded adaptation measures of farmers implicitly as climate induced. In fact some of them pointed out that farmers might undertake adaptive measures due to various motives, such as market and productivity (Maddison, 2006; Nhemachena & Hassan, 2007). Yet, they fail to distinguish these measures from climate induced adaptive strategies as this is a very difficult process and there has been no method of cross-checking whether farmers were actually responding to climate or non-climatic stimuli. Therefore, farmers' responses were taken for guaranteed as climate induced in the analysis of adaptive behavior. In such cases, analytical fatalities might be encountered where the analysis of determinants of adaptation might lead to erroneous conclusions due to the inclusion of non-climate driven adaptive strategies into the analysis.

Such limitations could be minimized, if not avoided, by bringing psychological factors into the central pot of the empirical analysis and devising some methodological remedies.

2.3.3.2. Applications Based on Protection Motivation Theory

In the abovementioned discussion, it is argued that the psychological aspect of adaptation has been ignored by the above approaches. In attempt to address this gap, Grothman and Patt (2005) developed a conceptual framework based on Protection Motivation Theory (PMT here after). Before, visiting their application, some time is devoted for discussing PMT, as this theory offers conceptual framework for this study's attempt to explain adaptive behavior of farmers.

Protection Motivation Theory is a psychosocial theory originally developed to explain protective behavior of people (prospectively and retrospectively) against health threats or risks. Besides health risks, the theory has been applied in a wide range of risk management issues like natural and technological hazards, biodiversity protection and online (internet) safety, and is found to be worthy in explaining people's protective behavior to the threats they perceive as risky (Menzel & Scarpa, 2005; Grothmann & Patt 2005; Martin et al, 2008; Bočkarjova et al, 2009).

According to PMT, individuals' engagement to protect themselves from a threat or harm is enhanced by four critical cognitions or perceptions (Block & Keller, 1998; Mulilis & Lippa, 1990; Floyd et al, 2000; Bočkarjova et al, 2009).

1. ***Perception of Vulnerability or susceptibility to the threat*:** the likelihood that the threat will affect the individual, or the individual's sense of susceptibility to hazards if no protective action is taken.

2. ***Perception of Severity of the threat***: the extent of harm the hazard would cause, or the degree of negative outcomes associated with the threat if there is no adaptive action.

3. ***Perception of Response efficacy:*** the belief that adaptive behavior works by safeguarding from the hazard, or the belief that protective action is effective in shielding oneself from the threat or reducing the threat.

4. ***Perception of Self efficacy:*** the perceived ability of an individual at performing or carrying out a protective behavior.

Furthermore, the theory assumes that rewards[33] and perceived barriers or costs (time, money, etc) influence the process of protective behavior (Pechmann et al, 2003; Norman, et al, 2005).

The theory organizes the above stated perceptual factors into two processes, namely threat appraisal and coping appraisal. While the perception of vulnerability and severity make up threat appraisal; response efficacy and self efficacy formulate coping appraisal (Armitage & Mark, 2000; Floyd et al, 2000; Menzel & Scarpa, 2005; Norman, et al, 2005; Bočkarjova et al, 2009). In threat appraisal, individuals assess their vulnerability to and severity of the risk before considering any adaptive behavior. This process results in some level of risk perception. In coping appraisal, they assess the effectiveness (response efficacy) of performing a possible preventive behavior and their abilities to carry out the behavior (self efficacy) taking into account the costs of performing preventive behavior. The result of going through these two appraisal processes leads to either adoption or neglect of preventive behavior (Bočkarjova et al, 2009).

Grothman and Patt (2005) molded this theory to fit into the context of climate change adaptation and developed a model which they named, model of private proactive

---

[33] Refers to internal and external incentives that might impede implementation of an adaptive behavior, or increase maladaptive response.

adaptation to climate change (MPPACC). Alike PMT, their basic assumption comprises two components: threat appraisal (with its resultant, risk perception) and adaptation appraisal (with its resultant, perceived adaptive capacity). The outcome of these two processes leads individuals whether to adapt (adaptation) or not to adapt. Adaptation will take place if risk perception and perceived adaptive capacity are high, or maladaptation[34] might occur when risk perception is high but perceived adaptive capacity is low. Whereas adaptation might not happen at all, when risk perception is low. With these constructs, they argue that it is only partially that objective adaptive capacity (socioeconomic resources) could determine adaptive behavior.

In their model, they depicted objective adaptive capacity as a direct determinant of adaptation and it represents basic resources (time, money, knowledge, entitlements, social and institutional support). They also supplemented the model by factors such as social discourse and adaptation incentives which are said to influence perception (risk perception and perceived adaptive capacity). Social discourse signifies the social and environmental context of the individual where risk perception and perceived adaptive capacity are influenced by the discourse of climate change (through media, interaction with friends, colleagues, neighbors and public agencies). They included this variable by taking the Social Amplification Theory of Risk into consideration, which posits the influence of broader societal, institutional and cultural context on hazard perception and response behavior (Kasperson et al, 1988). Adaptation incentive represents public and social support which motivates adaptation and could be represented by tax reductions, laws or social norms. Adaptation incentive mainly influences adaptation intention. Adaptation intention symbolizes those individuals who have intentions to adapt but do not carry out actual behavior due to some reasons.

Even though Grothman and Patt (2005) developed the model by including a number of complex cognitive factors that might affect perception irrationally, such as cognitive biases and heuristics, their interest in empirically testing the model was limited with risk perception and perceived adaptive capacity, as the two important components of adaptation process. They applied this model into two different settings: in Germany and Zimbabwe.

The first study was conducted on decisions of 157 randomly chosen residents on the banks of Rhine River in Cologne, Germany, to the threat of flooding. They run two

[34] According to Grothman and Patt (2005), maladaptation is an avoidant adaptive behavior where people evade actual adaptation process through avoidant reactions such as denial of threat and wishful thinking due to their low levels of objective means to respond, or carry out wrong response actions that rather increase damages. In their conceptualization, maladaptation is considered as an adaptive response where people react by denying or think wishfully to protect their psychological well being, even thought the responses (denial, wishful thinking etc) are not adaptive ones in the sense of preventing damage from a threat.

regression functions on four cases of residents' proactive flood prevention mechanisms against socio-cognitive factors (risk perception and perceived adaptive capacity), on the one hand, and socio-economic factors[35], on the other hand. Accordingly, socio-economic factors explained 3 - 35 percent of the variation in three of the cases and socio-cognitive factors did 26 - 45 percent of the variation in all of the four cases. The results show stronger influence of psychosocial variables on adaption decision of the residents than that of socio-economic factors.

The case in Zimbabwe was a qualitative one applied in a project setting where subsistence farmers were informed about seasonal climate variability (predictions of seasonal rainfall) so that they were expected to adjust their farming practices based on the information they had received. The observation showed that the farmers did not make any adaptation nor had an intention to adapt. They examined the reasons behind farmers' failure to adapt and found out that the farmers had perceived the risks not to be high and their adaptive capacity to be low, which is an evidence in favor of cognitive factors in determining adaptation.

In such way, Grothmann & Patt (2005) illustrated the importance of psychological factors which has not been addressed in other adaptation models in the context of climate variability and change. Even though, they demonstrated the importance of psychosocial variables in adaptation decision, they also raised a number of questions for further research. As to what extent the model could be applicable in various cultures and conditions, and in planned and aggregate adaptation decisions? What other socio-cognitive factors could be incorporated to enrich the model? How policy could influence cognitive factors to improve adaptive capacity? Since, their application of the model was on proactive adaptation, another question could also be added to the list of the above questions. To what extent does the model explain reactive adaptation? Answering this last question will be partly an attempt of this research. Another attempt will be to apply and evaluate the model among subsistence farmers in the Ethiopian context. In addition, as Grothmann & Patt (2005) did not specifically identify the contribution of each of the elements that make up risk perception and perceived adaptive capacity, this will also be another topic of concern to this research. Identifying the influence through the composites of risk perception and perceived adaptive capacity offers little to policy decision. Understanding the influence of each element helps to specifically target and address the factors according to their relative importance.

---

[35] Included age, gender, school degree, net income and housing tenure in their study.

Concluding Remark

Much of the discourse on climate change is going on at global and scientific level. The discourse has mainly knotted itself with scientific knowledge which has been largely supported by sophisticated computer programs. Even by doing so, it is not able to bring about consensus on the science, convince the skeptics and resolves uncertainties. In such world of knowledge, where debates will continue to fuel until scientific advancement prevails with convincing theories and economic and political interests are alienated from the reality of climate change, it is less likely that the world would converge in responding effectively to local concerns and vulnerabilities of the poor in the developing world. Until that level is achieved and the world converges in the fight against climate change, it is imperative that response actions at local levels have to be promoted based on local knowledge and indigenous experience. Even in the case of world consensus, it is very likely that the global discourse would channelize top-down policies that are less sensitive to localities and realities at the grassroots level. Either way, exploring and understanding local knowledge system will thus have paramount importance in designing policies that are responsive to local contexts. With this premise, this research underlines the quest for understanding climate variability and change at local levels through the lens of local knowledge and indigenous experience, and evaluate the authenticity of this knowledge by various methods. Addressing this constitutes part of the study.

Another issue discussed in this chapter goes with quantification of the impacts of climate change and variability on agriculture. Even though a number of economic models exist, limitations surrounding the models (besides uncertainties) made it hard to depict clear picture of the impacts. Since agronomic-economic models are more of experimental and expensive, Ricardian model is an appealing econometric model to several researchers as it uses observed (cross-section) data from actual farms. The central tenet of the Ricardian model is rooted on the assumption that land value, under well-functioning market, is the reflection of efficient use of land, which in turn, is influenced by climate conditions. Among other things, it assumes constant price, perfect market, costless adaptation process, and farmers as economic agents who are able to detect changes in climate so that they respond automatically. Nevertheless, some of its basic assumptions are non-existent in the developing world. Therefore, the application of the model among subsistence farmers (the interest of this study) appears to be problematic in particular and less applicable in the context of most developing countries like Ethiopia in general.

While quantification of the impacts among subsistence farmers yet requires innovative approaches that help us portray the intensity of impacts, understanding adaptation at micro-level is still another area which needs advanced research efforts. Pertaining to conceptual similarities with agricultural technology adoption, adaptation at micro-levels has been traditionally guided by agricultural adoption models. Based on this approach, a conceptual framework which states adaptation as a two step process has emerged and began stretching its roots in adaptation literature. Exclusively centering on resources, it tries to view adaptation as a function of socioeconomic and institutional (SEI) factors. The importance of SEI factors in facilitating and determining adaptation is beyond doubt. However, social science research clearly demonstrates the importance of subjective assessment of risk in one's decision to respond to threats. Psychological barriers could play an invisible firsthand role before SEI factors has even been conceived as determining factors. With this premise, paying due attention to the cognitive (psychological) aspect of adaptive behavior is another strand of this study.

Chapter Three

Theoretical and Methodological Framework

Introduction

In the previous chapter, the literature was examined and the gaps that this research tries to bridge are already identified. This chapter in turn outlines the theoretical framework and methodological approach that lead the study's attempt to address the gaps identified. The chapter entirely focuses and presents the general theoretical framework, the research approaches, site selection, sampling techniques and data collection instruments. Since the study is multidisciplinary, which brings together concepts and terms from geography, economics and psychology, it proposes a comprehensive theoretical framework that captures its multidisciplinary nature along with the main variables and concepts dealt in the study. The theoretical framework mainly sketches the overall linkage of the main questions of the study so as to pave the way for developing detailed analytical framework and models in the next consecutive chapters.

As the core concepts of the theoretical framework and of course, the study in general, the chapter begins with describing the terms climate variability and change, and spells out how they are operationalized in this study.

3.1. Conceptualizing Climate Variability and Climate Change

It is quite common to see the terms climate variability and climate change blended together in policy and research undertakings, predominantly in agriculture. Several research headings and literature bodies use both terms simultaneously without even mentioning as to whether what relates them together and where their boundary lays. Indeed, they are intermingled terms both in nature and in use, which made it very difficult to put clearly a visible demarcation line between them. That is why Preston & Stafford-Smith (2009) stated that the distinction between these two terms is largely artificial. Therefore, the distinction is barely noticeable, and arbitrary depending on aim and context. For the sake of operationalizing these interrelated terms, first their meaning and relationship are discussed and how they are used in this study is outlined at the end of this section.

In terms of definition, there are two major traditions of describing the term, climate change, although no agreed upon definition does exist. The first and relatively less popular definition regards climate change solely as an anthropogenic cause and climate variability as a natural cause (UNFCCC, 1992). The second and the most popular definition by IPCC regards climate change both as a cause of anthropogenic activities and natural processes, and hence climate variability is part and parcel of climate change. Other definitions usually emanate from these two authoritative definitions.

UNFCCC's definition stems from the central aim of the Convention, which aims to reduce anthropogenic greenhouse emissions in order to prevent dangerous interference with the climate system. Since UNFCCC is exclusively concerned with the reduction of greenhouse gases concentration in the atmosphere, it defines climate change strictly embedding its objectives.

> *Climate change means a change of climate which is attributed directly or indirectly to human activity that alters the composition of the global atmosphere and which is in addition to natural climate variability observed over comparable time periods (UNFCCC, 1992: Article 1, Number 2).*

On the other hand, IPCC, as a key scientific body, offers a definition based on its scientific outlook.

> *Climate change in IPCC usage refers to a change in the state of the climate that can be identified (e.g. using statistical tests) by changes in the mean and/or the variability of its properties, and that persists for an extended period, typically decades or longer. It refers to any change in climate over time, whether due to natural variability or as a result of human activity (IPCC, 2007a: 30).*

Unlike climate change, there are little differences, if any, in defining climate variability.

> *Climate variability refers to variations in the mean state and other statistics (such as standard deviations, the occurrence of extremes, etc.) of the climate on all spatial and temporal scales beyond that of individual weather events (IPCC, 2007a:79).*

Examples of climate variability include seasonal, annual, interannual, interdecadal and spatial variations in rainfall and temperature, extended droughts, floods, and conditions that result from periodic El Niño and La Niña events[36] (USAID, 2007).

---

[36] El Niño and La Niña are the opposite phases of ENSO (El Niño-Southern Oscillation) cycle. The ENSO refers to the fluctuations in temperature between the ocean and atmosphere in the east-central Equatorial Pacific. When ENSO deviates from the normal surface temperature, it results either in El Niño or La Niña. El Niño refers to "the large-scale ocean-atmosphere climate interaction linked to a periodic warming in sea surface temperatures across the central and east-central Equatorial Pacific". Whereas, La Niña signifies to "periods of below-average sea surface temperatures across the east-central Equatorial Pacific". (http://oceanservice.noaa.gov/facts/ninonina.html, accessed 03.03.2013).

Perhaps it is from IPCC's definition that researcher and policy experts' concomitant usage of the two terms originates from. In fact, other than the closely related definitions, the following two pragmatic relationships might enforce the less indivisibility of the two terms.

Physical relationship: Climate change, among other things, alters the pattern of temperature and precipitation, increases sea levels and most likely exacerbates the magnitude and frequency of extreme events (Sperling & Szekely, 2005; Kandji et al., 2006; Washington et al., 2006; IPCC, 2007c; World Bank, 2008). Most of these variables that climate change manifests itself are underlying elements that define climate variability (Kandji et al., 2006) and hence it is widely accepted that climate change alters the behavior of climate variability. Since climate change and climate variability are closely linked in the complex evolution of the climate system and climate change largely manifests itself with episodes observed in the current climate variability (Washington et al., 2006), it will be idealistic to deal with only one of them without paying attention to the other.

Quite usually climate change is regarded as future oriented in time and has global feature in space, and similarly climate variability exists at all temporal and spatial scales (as per the definition of climate variability). This overlap over space and time further broadens the complexity of the relationship between climate change and climate variability that arises from the natural coexistence of the two. Such complex relationships will continue to fuel the existing scholarly debate until the science of climate change matures enough to detect and attribute climate change accurately.

Impact and response relationship: As pointed out above, climate change partly and largely manifests itself with episodes observed in the current climate variability (Washington et al., 2006) but most likely with intensified magnitude and frequency (Kandji et a., 2006; IPCC, 2007c; Pittock, 2009). As such, distinguishing the specific impacts of climate change from those of climate variability is quite difficult (Burton et al, 2007).

On the other hand, the impacts of climate variability and associated management strategies are not new and are relatively well known than climate change. Thus, the existing impacts and response mechanisms to climate variability could be living laboratories to depict how climate change affects localities and what possible response strategies could be viable to the localities. That is why many researchers and policy makers stress that acquiring lessons and experience from impacts of current climate variability will be crucial to understand how climate change creates new vulnerabilities and exacerbates existing biophysical and social vulnerabilities and based on that formulate appropriate adaptation strategies, particularly at local levels (Eriksen, 2005; Sperling & Szekely, 2005; Washington et al., 2006; UNFCCC, 2007; NAS, 2010).

Table 3.1. The relationship between climate change and variability

Attribute	Climate Change	Climate Variability	Source
Cause	Man-made Could be also natural[37]	Natural in origin	UNFCCC (1992) IPCC (2007a)
Basic manifestation	Sustained shifts in the mean state of the climate or in its variability Exacerbation or inducement of climate variability	Variations in the mean state of climate and climate events Could contribute to climate change	IPCC (2007a) Rosenzweig & Hillel (2008)
Time scale	Long term (several decades, century)	Shorter term (monthly, seasonal, annual, interannual and interdecadal variation)	NAS (2010) Pittock, 2009 IPCC (2007a) USAID (2007) Burroughs (2007)
Spatial scale	Mainly global, but also could be regional and local	Mainly regional/local, but also could be global	
Response action	Mitigation/Adaptation	Mainly Adaptation Starting point for adapting to climate change	UNFCCC (1992) IPCC (2007a) Washington et al., (2006)

Source: Own synthesis based on the literature

Given such close association, it is quite evident why the two terms are concomitantly used in policy and research activities. In this current era, where isolation of climate change from climate variability is very difficult both in terms of its manifestations, temporal and spatial scales and impacts, and response actions of the later are considered as the first step to respond to the former, it remains more plausible to maintain the coexistence of the two in research and policy activities. Particularly, disentangling the impacts of climate change from climate variability in primary sectors such as agriculture, which is highly sensitive to climate, is largely problematic. Taking such closeness into account, climate variability and climate change in this study are approached concomitantly but literally referring to different phenomena of the climate.

Climate variability in this study is used to refer to short term climate variations and year-to-year fluctuation around the long term mean and in timing (e.g. delay/early onset, cessation of rainfall, interruptions) of the local climate. While this description has

[37] IPCC's definition; http://agroclimate.org/climate_change, accessed 11/01/2012.

a qualitative aspect, quantitatively it is depicted by annual mean values of the local climate and deviation from long term mean values (20 years). On the other hand, climate change is used to refer to continuous change (increasing or decreasing) in the state of the local climate in the past several decades, the trend in the past 2 decades, and projected climate values in the mid-21st century.

Since precipitation and temperature are the most essential climate factors for agriculture, these climate variables are used to explain climate variability and change in the localities covered by the study. In addition to their vital importance for agriculture, the limit of data in other variables of the climate forced the researcher to consider these climate variables only.

3.2. Theoretical Framework

As outlined in chapter 1, this study has three interrelated research themes. It is under these themes that it attempts to answer the research questions raised in section 1.3 of chapter 1. The themes are:

Theme 1: Exploring farmers' perception and experience of detection and attribution of climate variability and change along with impacts;
Theme 2: Analyzing the impacts of climate variability and change and adaptation with a specific focus on crop production and;
Theme 3: Investigating the socioeconomic and psychosocial factors that influence adaptive behavior.

A comprehensive understanding of climate variability and change in a developing country context unquestionably requires the social understanding of people's perceptions, impacts and adaptation practices. These core terms, i.e., perception, impact and adaptation are fundamental concepts that notably form the heart of social studies of climate variability and change at micro-levels. Through systematic linkage of the themes, an attempt is made to bring together these core concepts into a single pot along with the introduction of new concepts so as to better explain adaptation and sketch up a comprehensive visualization of impacts of climate variability and change in a vulnerable society at a time. Such aspiration, even though complex, requires a comprehensive theoretical framework which could commendably captures these core concepts inclusively. Theoretical frameworks applied in the assessment of vulnerability and adaptation to climate variability and change help to address this aspiration.

An understanding of impacts and adaptation in general emanates from the constructs of vulnerability. It is because of the existence of vulnerabilities of communities that we adamantly care to understand impacts and adaptation. Had it not been the characteristics that make societies vulnerable, we should not have paid due attention to impacts and adaptation at all. As such, the term vulnerability chains together the key research concepts in this research. Although vulnerability in the context of climate science has several definitions and approaches, a review of them is not an intention of this study, rather only focuses on contextualizing the term rooted within the process of assessing perception, impacts and adaptation.

Vulnerability[38] is context specific so that there cannot be a single and unifying definition and approach (Pearson & Langridge, 2008). Hence, stemming from the most authoritative and commonly quoted definition of IPCC (2001), the study builds a comprehensive theoretical framework applicable in the context of subsistence farming communities. Vulnerability, in the context of climate variability and change, is defined as "the degree to which a system is susceptible to, or unable to cope with, adverse effects of climate change, including climate variability and extremes" (IPCC, 2001:995). As a functional form, vulnerability is depicted somewhat consistently in the literature as a function of exposure, sensitivity and adaptive capacity of that system (Smit & Wandel, 2006). IPCC (2001) further defines these variables as follows. Exposure is the nature and degree to which a system is exposed to significant climatic variations. The degree to which a system is affected by the exposure to climate-stimuli either adversely or favorably defines sensitivity. The higher the degree of these two variables, the higher is the level of vulnerability. Conversely, adaptive capacity counteracts the level of vulnerability. Adaptive capacity is the ability or capacity of a system to adjust itself to climate stimuli to moderate potential damages, to take advantage of opportunities, or to cope with the consequences. Brooks & Adger (2005:168) define adaptive capacity in concrete terms as "the ability to design and implement effective adaptation strategies, or to react to evolving hazards and stresses so as to reduce the likelihood of the occurrence and/or the magnitude of harmful outcomes resulting from climate-related hazards". This study adopts this definition of Brooks and Adger's (2005) as it more qualifies for practical and analytical purposes.

---

[38] It has to be noted that the term vulnerability in this research is not a core concept but mainly used to chain the key concepts involved in the research. In fact, vulnerability is a broad term which entertains various approaches. In the literature, there are three broad approaches of vulnerability: risk-hazard approach, social constructivist approach and integrated approach (Füssel & Klein 2006). The risk-hazard approach is an approach commonly applied in natural hazards and focuses on the biophysical damages (economic, monetary, human causalities, mortality and ecosystem damages) as a result of hazard or climate stimuli (external cause) (Brooks, 2003; Füssel & Klein 2006; Cutter et al., 2009; Ribot, 2009). The social constructionist framework, on the other hand, focuses on characterizing the internal social, institutional and political causes that make people vulnerable in the first place (Adger & Kelly, 1999). An integrated approach combines these two approaches to understand vulnerability of a system (Cutter et al., 2009; Ribot, 2009). The framework outlined in this study, i.e., vulnerability as a function of exposure, sensitivity and adaptive capacity belongs to this last approach, where exposure refers to external dimension (climate variability and change) and the last two elements imply an internal dimension of a system (Füssel & Klein 2006). Beyond this, there are also other approaches, such as sustainable livelihoods approach which is mainly used as a tool in development planning and programs.

In this study, the element of exposure is the socioeconomic systems of farming communities in general and farming system (crop) in particular. These systems confront with variations of the most basic climate variables, i.e. precipitation and temperature. As indicated earlier, due to their vital importance for crop production and of data limit, these two climate variables (here after referred to as climate stimuli/climate variables unless referred individually) are used to designate the degree of exposure of the systems to climate variability and change. The types of communities covered in this study are rural households whose livelihoods are dependent on farming. Since their livelihoods are entirely dependent on farming (especially crop production), this sector will be of a particular interest in this framework of analysis.

The nature of the farming is subsistence. Subsistence agriculture is characterized by dependence on family labor, labor intensification because of its surplus nature and use of rudimentary tools, small and fragmented land holdings, no or little application of external inputs and technology (fertilizer, improved seed etc), and low productivity (Barnett et al., 1996; Ellis; 2000; Heidhues & Brüntrup, 2003). On top of this, farming is dependent on the amount and distribution of rainfall, which is a typical characteristic of Ethiopian agriculture. These characteristics make subsistence agriculture sensitive even to little fluctuations in the climate variables. Furthermore, social vulnerability which has been inherited in the socioeconomic and political systems exacerbates the sensitivity of crop farming. Typical examples of social vulnerability among subsistence farmers could include lack of access to institutional support, infrastructure, livelihoods' diversification and higher demographic dependency ratio.

The relationship of sensitivity with the level of exposure of the farming to climate stimuli defines the impacts on crop farming (Theme 2a, see Fig. 3.1 below). Farmers are expected to react in some way to reduce the impacts. For that, first they have to be aware of the signals of climate stimuli, and then assess the expected level of susceptibility of the farming and severity of the exposure to climate stimuli, and evaluate their adaptive capacities. As discussed in the previous chapter, this perceptual process is a decisive step to any action of adaptation and is partly depicted by Theme 1. Adaptive capacity by itself is not adaptation. But it is a spectrum of socioeconomic and institutional enabling and disabling conditions for adaptation. Borrowing physics terms, it could be portrayed as a potential energy where adaptation as a kinetic energy evolves from. Therefore, "adaptations are manifestations of adaptive capacity" (Smit and Wandel, 2006:287). Adaptive capacity comprises of a set of resources which are clearly described by Brooks and Adger (2005:168).

> *The adaptive capacity inherent in a system represents the set of resources available for adaptation, as well as the ability or capacity of that system to use these resources effectively in the pursuit of adaptation. Such resources may be natural, financial, institutional or human, and might include access to ecosystems, information, expertise, and social networks.*

Throughout the literature, it is also customary to link adaptive capacity to natural, economic, social, institutional and political factors (Brooks & Adger, 2005; IPCC, 2007a; Petermann, 2008; Schlickenrieder et al., 2011). These factors represent objective adaptive capacity. But, the existence of objective adaptive capacity does not guarantee adaptation (Levina, 2007). In line with this, IPCC (2007b:719) asserts that even having "higher adaptive capacity does not necessarily translate into actions". The above quotation (Brooks and Adger, 2005) also indicates the necessity for a capacity that utilizes the resources into adaptation. Enabling conditions such as political will and facilitative institutional systems must be there (Ibid). However, at individual level, particularly where resources are meager, persuasive political and formal institutional systems are lacking and external support is minimal, psychological or perceptual (cognitive) variables are very important to translate adaptive capacity into adaptation. That is why IPCC (2007b) emphasizes cognitive and psychological factors as important factors for adaptation along with other socioeconomic factors. Adaptation here refers to concrete and specific actions carried out by farmers as a response to climate stimuli. As argued in chapter 2, the psychological (perceptual) variables of PMT, namely, perception of susceptibility (perceived vulnerability), perception of severity, adaptive efficacy, self-efficacy along with tangible resources constituent the overall adaptive capacity. As tangible resources such as finance and technology represent objective adaptive capacity, perceptual variables on the other hand represent subjective adaptive capacity. Adaptive capacity is thus composed of two types of capacities as depicted in Fig. 3.1. Such conceptualization helps to integrate objective and subjective adaptive capacities that are inherent to actual adaptation processes.

Subjective adaptive capacity plays two roles as a process and a channel of linkage between climate stimuli/exposure and adaptation. As a link, it involves detection and attribution of climate stimuli, and gives rise for the assessment of risks and impacts and based on the assessment it helps to initiate the quest for adaptation (Theme 1). As a process, it involves the decision process as to whether farmers decide to translate objective adaptive capacity into adaptation (Theme 3a). While a functional combination of perception variables could show the overall level of subjective adaptive capacity, each variable, on the other hand, plays a role in adaptation decision. It is assumed that farmers assess the risks of climate stimuli through perceived vulnerability and perceived severity. Given some level of risk threshold, they pass to evaluate the effectiveness of performing possible adaptation option (adaptive efficacy) and their abilities to carry out the option (self efficacy) taking the costs of adaptation (perceived cost) into account. The final result of this process leads farmers to some sort of adaptation decision (either to adapt or not to adapt). As said earlier, this is particularly true among subsistence farmers where there is very limited access to resources and external support. As such psychological variables (subjective adaptive capacity) play an important role in adaptation decision (Theme 3a).

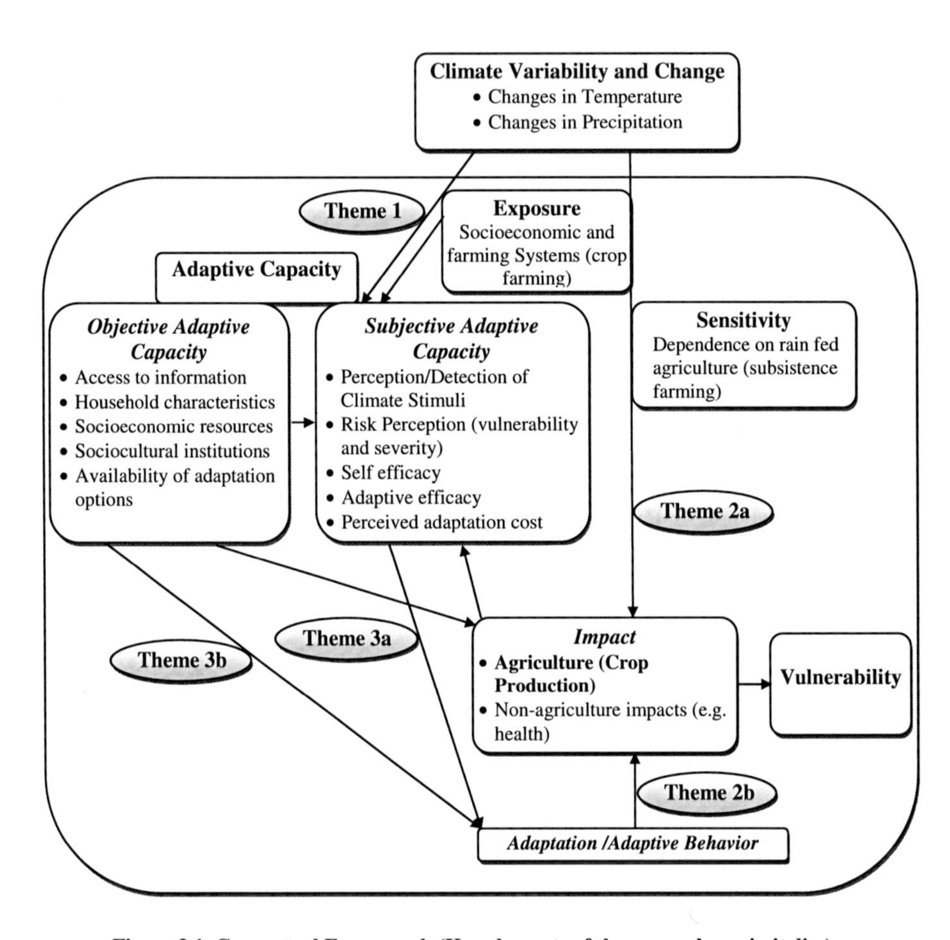

Figure 3.1. Conceptual Framework (Key elements of the research are in italics)

Objective adaptive capacity, as depicted in Theme 3b, includes SEI variables and directly determines adaptive behavior of farmers provided that the psychological variables have produced the motivation required to adapt. Assuming that farmers are rational in their decision to respond to climate stimuli, they will exhibit adaptive behavior (performing concrete adaptation action) given some level or threshold of both subjective and objective adaptive capacity. They adapt to reduce the impacts on crop production and minimize future vulnerabilities of farming to unfavorable climate stimuli. Therefore, adaptation counteracts the impacts of climate stimuli (Theme 2b).

The combined assessment of the climate impacts (Theme 2a) and adaptation (Theme 2b) would show the overall impacts on crop production. But, farmers' livelihoods are still subject to social vulnerability that cannot be addressed by climate specific measures (adaptation). Consequently, the seepage of impacts continues to exist as a form of an outcome or residual vulnerability[39]. Such seepage of impacts is further stimulated by the limits of adaptation that cannot restrain all levels of farming's exposure to climate stimuli. The final impact of climate variability and change is thus captured at the level of outcome vulnerability, where estimations from regression analysis depict a figure for interpretation.

This theoretical framework is presented to show the general relationship of the core concepts of the study, and seeks to illustrate the broad conceptual framework in operationalizing the research. Nevertheless, this theoretical framework will be broken down to various analytical frameworks in the next chapters to address each of the specific themes piece by piece. It has to be noted that this framework centers exclusively on farming communities, since they are the major subjects of the study (to analyze local knowledge, adaptive behavior, and impacts). Yet, the study uses external sources of information, such as experts and meteorological data. It employs these sources only as reference points to verify farmers' perceptions and knowledge and to use as inputs in operationalizing the framework, and hence no special emphasis is given in the framework but they are covered by specific analytical methods in the coming chapters.

3.3. The Research Approach

Realization of the objectives of the interrelated themes of the study entails mixed methods approach, which combines both qualitative and quantitative methods across the stages of the research process, and integrates them essentially in the process of data collection, analysis and interpretation. Although mixed methods design pulls in complexity in the overall process of the research process, it forms a fertile ground to exploit the advantages of both qualitative and quantitative methods while minimizing the weaknesses of exclusively relying on one method. Such exploitation of the methods enhances the integrity of research findings and helps to augment findings from different angles, and creates a comprehensive account of the research area (Bryman, 2006).

Given the five purposes of applying mixed methods design as exemplified by Greene et al (1989), the rationale for using mixed methods in this research stems both from its advantages and the intention to address the multifaceted themes of the research through various forms of data collection, data analysis and interpretation techniques. Greene et al (1989) indentified five purposes of applying mixed methods design.

---

[39] Refers to a form of vulnerability which sustains after counteractive measures (such as adaptation) are considered (Jones et al., 2007; IPCC, 2007b).

a. Triangulation: a design that employs different and multiple data collection methods with intent of better understand the research problem by corroborating, converging and corresponding evidences.
b. Complementarity: seeks to elaborate, enhance, illustrate and clarify the results of one method with the results obtained from the other method.
c. Development: wherein the researcher uses the results obtained from a preceding method to construct and shape subsequent methods (e.g. instrumentation, sampling, analysis strategies) that the researcher seeks to employ.
d. Initiation: is a rare practice where new inquires, unanticipated results and contradictions from one method recast and instigate new lines of thinking and pose challenges to the original conceptual framework of the study so that other methods are employed to explore and enrich these new insights.
e. Expansion: as the name itself tells, here the researcher widens and intensifies the scope of the research problem through the application of different methods.

This research mainly exploits the first two purposes of mixed methods design. As such, the research employs the convergent parallel design where both qualitative and quantitative data are collected concurrently, analyzed separately, and results from both approaches (qualitative and quantitative) are compared and contrasted during the interpretation process. Since the study is partly based on subjective assessments (perception), multiple data collection methods were employed so as to cross-examine evidences from different viewpoints (sources) and complement the results. As such, through the application of various data collection methods, detailed documentation was constructed on detection and attribution, socioeconomic impacts, stress factors from farmers' perspectives as well as experts' views in some of relevant issues. Qualitative methods were utilized to explore subjective assessment about climate variability and change (documentation on detection, attribution, details of impacts, climatic and non-climatic stress factors, and barriers to adaptation). On the other hand, quantitative methods were employed to triangulate and complement qualitative evidences through objective data (meteorological recordings and household's socioeconomic data) and further elaborate the impacts of climate variability and change focusing on crop production and associated adaptive behavior of farmers.

In the first part of the research, both qualitative and quantitative methods were employed to describe and explore the trends and characteristics of climate variables in the study areas based on farmers' subjective assessments and meteorological recordings obtained from the respective weather stations. More specifically, cross-examination was carried out as to whether the results from subjective assessment on the one hand and results from objective assessment on the other hand could converge or diverge in telling the stories of local climate. This was further augmented by farmers and experts' description of the climate, and comparing and contrasting their views on detection and attribution of climate variability and change based on practical illustrations that farmers encounter in their day-to-day activities. Associated socioeconomic impacts were also analyzed based on farmers' own observation. Since vulnerabilities (impacts) caused by climate factors could be disguised with non-climate factors, further scrutiny was carried out to indentify non-climatic stress factors that are believed to be serious sources of concern for farmers' livelihoods. The purpose of investigating the stress factors is to help clarify the importance of climate factors as stressors in the face of other multiple stress factors.

In the second part, following a broad description of the socioeconomic impacts of climate variability and change documented in the first part of the analysis, data were reduced with an intention to elaborate, illustrate and simulate the impacts with a closer look into one single fundamental part of farmers' livelihoods (crop production) through quantitative analysis. In the interpretation process, the results obtained from this analysis were compared and contrasted with the results from qualitative analysis (obtained from the first part of the analysis).

In the third and final part of the study, adaptive strategies that have been performed by farmers were described along with the barriers that impede them. Finally, a thorough investigation was carried out to identify the factors that influence adaptive behavior using theoretical triangulation (from psychological and socioeconomic perspectives).

3. 4. The Study Area and the Selection Process

Ethiopia is an agrarian country located in the Horn of Africa between 3^0 to 18^0 N latitudes and 33^0 to 48^0 E longitudes. It has the land size of about 1.12 million km^2 (NMSA, 2001:1). According to the 2007 census, its population amounts to 73, 918,505 with annual growth rate of 2.6 percent (CSA, 2008:1, 11). Of these, about 84 percent lives in rural areas (CSA, 2008:19).

The country is well known for its topographic diversity which comprises high mountains, rugged terrain, flat topped plateaux, deep gorges, river valleys and plains (Plant Genetic Resources Center, 1995). Very commonly, the country's landmass is classified into two as highlands and lowlands based on altitude. The land area above 1500 meters from sea level comes under highlands and covers about 45 percent of the country's land size (NMSA, 2001:30). Pertaining to conducive environmental conditions for settlement and agriculture, the highlands are home to 90 percent of the country's population (Kloos & Adugna, 1989). Although, the low lands (areas lying below 1500 meters) account slightly higher size of the country's land, the size of inhabitants is very low due to the prevalence of tropical diseases, low levels of rainfall and moisture, higher temperature and shallow soil (NMSA, 2001; ERM, 2003; Worku, 2007). The dominant agricultural activity in these areas is mainly livestock rearing carried out by pastoralists (NMSA, 2001; Mengistu, 2003).

From the perspectives of river drainage systems, the country could be classified into four major basins (FAO, 2005).

a. **The Nile Basin**: the drainage system which stretches from the northern and central parts to westwards and accounts about 33 percent of the country's land area. It includes the Blue Nile (Abay) Basin, Baro-Akobo, Setit-Tekeze/Atbara and Mereb basins.

b. **The Rift Valley**: this includes river and lake drainage systems which includes Awash River, Omo River and Lakes Basin and covers about 28 percent of the country's land area.

c. **The Shebelli-Juba Basin**: this basin drains the southeastern mountains and lowland areas and covers about 33 percent of the country.

d. **The North-East Coast**: this basin widens to east coastal areas towards the Gulf of Aden and covers 6 percent of the country.

The other form of classification[40] of the country is usually carried out based on climate which is justified by climate's predominant role in determining population settlement and agricultural activities in the country. These classifications include the traditional system, Köppen's classification, Thornthwaite's classification (moisture index), and Agroecological Zone (AEZ) (Gonfa, 1996; CSA et al., 2006). The traditional classification further splits the highland-lowland classification and subdivides Ethiopia into 5 climate zones similarly based on altitude (EPA, 1998; NMSA, 2001; http://www.fas.usda.gov/pecad2/highlights/2002/10/ethiopia/baseline/Eth_Agroeco_Zones.htm, accessed 23/05/2012). These are:

a. **Wurch (cold highland)**: covers highland areas above 3,000 meters and mainly barely is grown. In addition, it is primarily used for grazing.

b. **Dega (cool, humid highland)**: a highland in between 2,500 and 3,000 meters and crops like barely, wheat, oilseeds and pulses are grown.

c. **Weyna Dega (temperate, cool sub-humid highland)**: this is a highland area in between 1,500 and 2,500 meters. In this zone, the major crops include teff, wheat, barley, maize, sorghum and chickpeas.

d. **Kolla (warm, semi-arid lowland)**: comprises the lowlands in between 500 and 1, 500 meters. The principal crops include sorghum, finger millet, sesame, cowpeas, and groundnuts.

---

[40] For detailed classification see NMSA (1996); Gonfa (1996); CSA et al (2006).

e. **Bereha (hot and hyper-ardi lowland)**: represents the hot lowlands below 500 meters above sea level. Here, crop production is very limited.

Based on moisture index, one can also find the categorization of the country into 5 zones as humid, moist sub-humid, dry sub-humid, semi-arid and arid (Gonfa, 1996). In Köppen's classification, which roots on precipitation and temperature, the country could be classified into three major climate zones as dry climate, tropical rainy climate and temperate rainy climate. These three climatic categories are also subdivided further into 11 climate zones which could range from equatorial desert to cool highlands. In this classification, one has to be cautious that the temperate climate in Ethiopia is not truly temperate but symbolizes merely a high altitude features (NMSA, 1996). Another detailed classification developed by the Ministry of Agriculture and Rural Development redefines the country into 18 major agro ecological zones (CSA et al, 2006). It mainly restructures the moisture index classification into detailed subsections. However, the traditional classification as a classical system and Agro ecological zone as a detailed classification system are quite widely used and helpful for understanding land use and agricultural practices in the country (Coppock; 1994; CSA et al., 2006).

Consulting these classification vis-à-vis the dominance of population settlements and agricultural activities, two major geographic locations from the drainage basins, namely Nile Basin and Rift Valley were considered for the study. These areas together account for about 61 percent of the country's land size and have higher share of agricultural activities. From these basins, Blue Nile (also called Abay Basin in Ethiopia) and Rift Valley Lakes Basins were picked because of their importance in agricultural production. For instance, Blue Nile Basin alone accounts about 40 percent of the national agriculture production (http://www.mowr.gov.et/index.php?pagenum=3.1, accessed 24/05/2012). The location of these basins in different but biggest regional states[41], and while one is river basin and the other is lakes basin[42] is another opportunity that come to emerge from this selection.

Once these areas were picked, the next step was to delimit the basins in order to focus on specific sites for detailed analysis. In this case, one can randomly select the sites. However, the probability of ending up anywhere in sparsely populated areas or pastoralist localities where crop production is nearly negligible, or in lowlands (for

[41] From regional perspectives, one can also argue with the plausibility of these areas by representing the two largest regional states of the country in major demographic and socioeconomic settings such as population size, land area and agricultural production (BNB from Amhara National Regional State, and RVLB from Oromia National Regional State). Oromia National Regional State and Amhara National Regional State account 36.7 percent and 23.3 percent of the country's population, respectively (CSA, 2008). They are the 1st and the 2nd in the country in terms of population size.

[42] The selection of lakes basin is an added asset as this had created a study opportunity by covering areas of lakes basin along with that of river basin, as compared to abundant and moderately studied river basins in the country.

instance, Bereha, where there is limited crop production) cautioned for the consideration of at least some loose criteria. Hence, areas within dominant climatic zone (both in terms of settlement and agricultural activities) and that grow major grain crops were considered. In this case, Weyna Dega (classified under highland) is most dominant agricultural zone where major rainfed crops are grown (Hurni, 1998). Weyna Dega is also dominant in its population size where about 75 percent of the country's population lives in (Bekele & Hailemariam, 2010; http://countrystudies.us/ethiopia/43.htm).

The final exercise of the criteria led to the selection of two Weredas[43] from the Blue Nile Basin (here after, BNB) and three Weredas from Rift Valley Lakes Basin (here after, RVLB) as the physical destinations of this research.

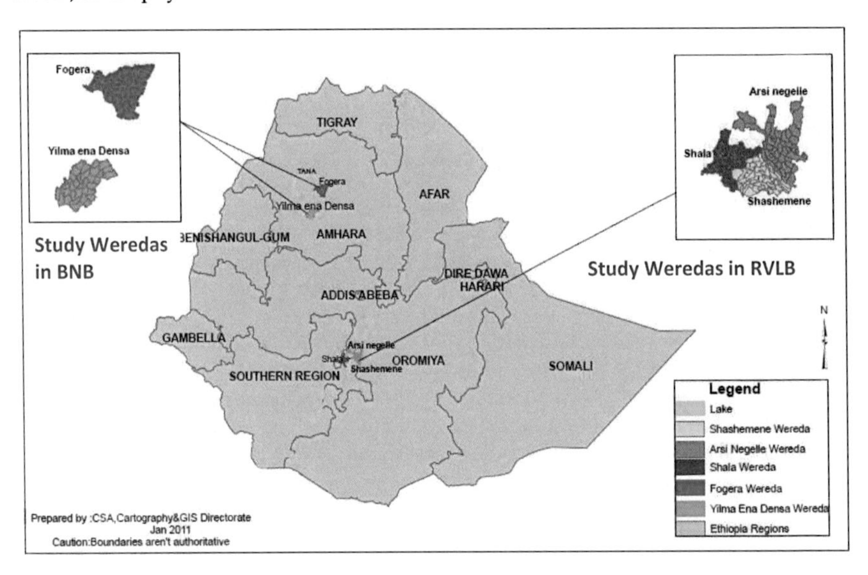

Figure 3.2. Geographic location of the study areas

At first, two Weredas from each Basin were thought. Later after releasing the existence of some interruptions in climate recordings of weather stations in RVLB, one more adjacent Wereda was included in this area so as to minimize the problem that might arise due to lack of climate data (particularly in trend analysis). Indeed, missing data was not a problem when it comes to regression analysis in estimating the impacts of climate on crop production as weather data for that particular year (2009) was available.

[43] Wereda is a provincial administration which is comprised of lowest level administration called Kebele. Kebele in turn is the lowest administrative district which comprises of several villages.

Finally, Yilmana Densa and Fogera Weredas from BNB, and Shala, Arsi Negele and Shashemene Zuria Weredas from RVLB became part of the study. Under traditional climate classification, these Weredas belong to Weyna Dega (highland) but differ in agro ecological zonation, where Yilmana Densa and Fogera come under Tepid to Cool Moist Mid Highland and Shala, Arsi Negele and Shashemene Zuria under Tepid to Cool Semi-Arid Mid-Highland. A total of 8 Kebeles each comprised of several villages were included in the study.

3.5. Sampling and Data Collection Process

The study was conducted based on both primary and secondary data obtained from various sources. Primary data was collected from farming households, experts and farmers, and focus group discussions. Household characteristics, socioeconomic and psychosocial information, production inputs and outputs, and adaptation strategies were collected from sample households in each of the Kebeles with the help of semi structured questionnaire. Simple random sampling was used to select the households from the Kebeles for filling the semi-structured questionnaire. A total of 250 households, 125 from each of the basins, were interviewed by enumerators. The sampling size was decided based on time and financial constraint, and the nature of households' spatial distribution, where houses are scattered over space in far distances to each other.

Focus group discussions were also carried out with farmers to explore local knowledge about historical accounts of the climate and agricultural practices in their respective areas. Such type of information by nature requires elderly farmers who have deep knowledge of agricultural and ecological characteristics of the localities. Consequently, elderly and experienced farmers were selected purposefully by the help of agricultural development agents and Kebele administrators. Two focus group discussions comprised of 5 to 9 people in each of the basins were undertaken. An in-depth discussion was made on issues such as agricultural activities, the overall ecological behavior of their surroundings, characteristics and trends of precipitation and warmness and coldness of days, impacts of climatic stimuli and other factors that affect their livelihoods.

Interview was conducted with farmers and experts. Each Kebele has at least three specialized agricultural development agents who are deployed by the government. Each of the agents is responsible for crop, livestock and natural resources management, respectively. Expert interviews mainly included these experts and in some cases environmental and health experts at regional and/or Wereda levels.

An in-depth interview with women and men farmers was also conducted with the aim of enriching information collected from focus group discussions and documenting personal accounts, particularly on how they detect changes in the climate and to what possible factors they attribute the changes they perceived. The interviewees were selected by agricultural development agents and Kebele administrators based on the interviewee's background knowledge on local agro ecological issues, and experience in beverage preparation (in particular to women).

On the other hand, secondary data was obtained from National Meteorological Agency of Ethiopia (NMA), Central Statistics Agency (CSA), agricultural and health offices at various levels, and Werede and Kebele administrations. The National Meteorological Agency of Ethiopia (NMA) is the sole source of climate recordings (precipitation and temperature). Data for precipitation and temperature were obtained for each of the weather stations representing the study sites along with adjacent stations.

A problem of missing data and absence of recordings was encountered mainly in RVLB, and such problem was not observed in BNB with the exception of few missing values. Unless prolonged absence of data was encountered, infrequent missing values were filled using inverse distance[44] estimation method. This method was used because some adjacent weather stations near to the station with missing data are available so that relatively representative figures could be estimated as it (the method) exploits the proximity of the weather stations. In cases, where estimation through inverse distance method was not possible due to prolonged absence of recordings, data from the nearest weather stations was considered. Similarly, as there was no recordings of temperature for Shashemene weather station, the nearest neighboring weather station (Awassa) within 20 kilometers was considered. In the case of Aje (Shalla Wereda) where it suffers from poor recordings, Alaba Kulito's recordings were used additionally. In fact Awassa alone as a nearby weather station in southern direction from Aje could also serve the station. But to enhance the quality of results, Alaba Kulito as a nearby station within 30 kilometers to the west of Aje was also used along with Awassa weather station. Using

---

[44] Inverse distance method is an estimation technique of a missed weather value of a station which is given by:

$$P_x = \frac{\sum_{i=1}^{N} \frac{1}{d^2} p_i}{\sum_{i=1}^{N} \frac{1}{d^2}}$$

where, P_x is the estimate rainfall value of a weather station with missing precipitation; $p_i$the rainfall value of surrounding weather stations; d is the distance from each of the weather stations to the weather station with missing rainfall value, N is the number of surrounding weather stations (De Silva et al., 2007). This method is one of most used techniques to interpolate climate data as well as extensively used in the mining industry (Collines & Bolstad, 1996, Hartkamp et al., 1999; Legates & Willmott (1990) & Stallings et al (1992) in Luo et al., 2008). It is advantageous to utilize proximity of adjacent weather stations with data to a weather station with missing data. The method has also shown to work well with noisy data as well (Collines & Bolstad, 1996).

the nearby stations in these cases not only helps to fill the data gap but also to demonstrate the bigger picture of the climate trend in the surrounding areas and cross-check the trends of climate in the nearby weather stations. As a matter of fact, such consideration of the nearby weather stations was used only for trend analysis of the climate variables. For impact analysis (in cross-section data), both precipitation and temperature data from the respective weather stations were used as data in this regard was available in all of the respective weather stations with the exception of Shashemene's temperature recordings, where grid data of Shashemene itself was used.

To forecast the impacts of climate change, downscaled climate change data (precipitation, maximum and minimum temperature) was obtained from the World Bank's Climate Change Knowledge Portal. The Portal provides downscaled data for Ethiopia and other countries. The mid-21st century data that runs from 2046-2069 under greenhouse emission scenario of A1B was considered. This period was chosen mainly due to data availability (at the time of data extraction) in the nearest possible period to the current era. Considering nearest periods to the current era apparently reduces uncertainties in the predicted climate data.

3.6. The Design of Questionnaire and Interview Procedure

One of the contributions of this research centers on its methodological approach in eliciting farmers' adaptation strategies. As discussed in the previous chapter (section 2.3), several studies had drawn adaptation strategies that were carried out by farmers just by asking them (farmers) what adaption strategies they had made as a response to climate change (Maddison, 2006). Normally, such approach might lead farmers to respond in a certain way and hence influence the authenticity of farmers' responses in several ways. First, it might lead farmers to answer in such a way that they believe what the interviewer wants to hear about. Second, they might assume themselves as if they were supposed to undertake adaptation strategies and might respond positively as if they were undertaking while not in actual cases. Third, they might also tag adaptation strategies that they had adopted as a response to non-climate causes as if they were performing as a response to climate change. In this case, non-climate responses might be regarded as climate responses and thus might be wrongly included into the analysis. To minimize the possibility of such bias, if not avoid completely, the questionnaire was designed in a systematic way as outlined below and the same approach was followed in the interview process.

By consulting the literature, a list of adaptation strategies was developed in the first place. This list was then checked against local practices with a close consultation of experts, and then nine adaptation strategies regarded as most widely practiced in the localities as a response to climate variability and change were entered into the final questionnaire. In the first part of the questionnaire, respondents were cross-checked against each of the strategies as to whether they were performing (or not) while without mentioning about climate variability and change. Later at the last part of the questionnaire, where it seeks information about climatic conditions and adaptation, farmers were asked as to whether they have performed adaptive strategy (strategies) as a response to climatic problems and perceived trends of the climate which they expressed in the preceding questions.

Through such cross-checking of the first part of farmers' responses (application of adaptation strategies) with the last part of the questionnaire (about climatic conditions and adaptation) an attempt was made to identify as to whether farmers were performing adaptation strategies as response to climate or non-climate reasons. It was assumed that while a farmer is performing an adaptation strategy and if he/she says that he/she is not adapting or responding to climate variability and change (as checked at the later stage of the questionnaire), this might be an indication that the farmers' performance of the strategy is most probably for non-climatic reason(s). In such way, extra scrutiny was carried out to identify those farmers who have exactly applied adaptation strategies for the purpose of climatic reasons.

Similarly, the interview process (in particular with women) was organized systematically to grasp responses from the respondents without being biased by the subject under discussion. Rural communities, especially women, have less or no access to weather recordings. Their knowledge about the climate and its trend is mainly built based on experiential learning which in turn develops from day-to-day activities associated with their livelihoods. To understand these experiential learning, an in depth interview was carried out. In this interview, a series of questions about food preparation and preservation practices were presented for women in particular and they were made to state the length of food and beverage fermentation and preservation trends of the present with respect to the deep past, like 20 or 30 years ago. Following their reaction to the above inquires (if they said that changes have happened), they were also asked as to why changes in the length of food fermentation and preservation had occurred. In interviewing, care was taken to keep the authentic reactions and views of respondents by posing the questions tacitly without mentioning about climate change and variability.

3. 7. Data Analysis

The multidisciplinary nature of the research made it difficult to describe the detailed methods of data analysis right here in this section. Separately elaborating detailed methods of analysis for each of the themes is believed to be more straightforward for embedding the methods in particular context of the themes. Consequently, the research followed different courses of analysis in addressing each of the specific themes. Therefore, the methods are detailed in the consecutive chapters that deal with each of the themes separately. Nevertheless, a general picture of the methods of data analysis could be depicted briefly here under.

In the first place, interview notes and transcribed focus group discussions were summarized, and data obtained from questionnaire were entered into statistical software. Subjective assessments of farmers and experts (perceptions, detection, attribution, impacts and characterization of the long term climate) were analyzed descriptively and key themes were identified from the summary of the transcribed data. Content analysis was employed to identify key themes, characteristics of the climate in the past several decades, impacts that are associated with the climate and attribution of climate variability and change. Major ideas and views obtained from characterization of the climate, impacts and attribution of climate variability and change were illustrated further by direct quotes and personal accounts of respondents.

Comparative analysis (triangulation) was made by comparing and contrasting subjective assessments with objective data (meteorological data) on the trend of the climate, farmers' views with that of experts' on attribution of climate variability and change, and climate factors with that of non-climate to identify the major source of concern among farmers. Once general characteristics of the climate and its trend, detection and attribution experiences, impacts and major sources of concerns were identified and explored, the next analysis focused on application of econometric models to estimate impacts and identify the factors that influence adaptive behavior. Multiple regression was used to estimate the impacts of climate variability and change and adaptation on crop production. Logistic regression, as a qualitative choice model, was also employed to predict adaptive behavior. In so doing, various statistical and model diagnostic tests were carried out to address statistical and econometric issues.

Concluding Remark

This chapter has presented the theoretical framework and methodological approach of the study. Given the study's three interrelated themes, it proposed a comprehensive theoretical framework that reflects the multidisciplinary nature of the research with due consideration to link the themes and core concepts dealt in the study. Methodologically, it involved mixed methods approach that combines both qualitative and quantitative approaches across the stages of the research process. The comprehensive nature of the study entailed the term vulnerability as an interconnecting term of the core concepts of the study.

Chapter Four

Exploring Local Knowledge on Climate Variability and Change

Introduction

Traditionally, climate variability and climate change have been understood and analyzed based on climate recordings. Leading documents and IPCC publications all attempt to assert their positions on the terms solely based on these recordings. With the acknowledgement of local (indigenous) knowledge in pushing further the frontiers of our understanding on the topic, specially at local levels, IPCC is expected to include a distinct topic that deals with local knowledge and specific circumstances of vulnerable groups in the coming Fifth Assessment Report (AR5)[45] (Nakashima et al., 2012). Therefore, studying local knowledge is yet an emerging field of research in the context of climate change and variability.

This chapter is therefore devoted to explore this knowledge system (local knowledge) in the context of subsistence farming communities. Local knowledge here refers to farmers' understanding about the state of local climate that they developed from experience and observation, and arrays of information transferred through generations. Following some scientific steps involved in the study of climate variability and change, this chapter specifically attempts to document this knowledge based on farmers' perception, detection and attribution experience. Besides this, it further looks at the impacts through the lens of farmers' knowledge, and concludes whether the impacts are likely climate induced or not. In so doing, it verifies farmers' knowledge in light of meteorological records and experts' views.

4.1. Analytical Approach

As indicated in the introductory part, this chapter addresses the issues raised under Theme 1 of the study's theoretical framework, as highlighted in the figure below. Towards this end, it principally adopted the analogy of the scientific study of climate variability and change.

---

[45] IPCC's Fifth Assessment Report is expected to be approved stage by stage (reports from the three working groups and the synthesis report) in 2013/14.
(http://www.ipcc.ch/activities/activities.shtml#.UDSO9dbibjI, accessed 02/09/2012).

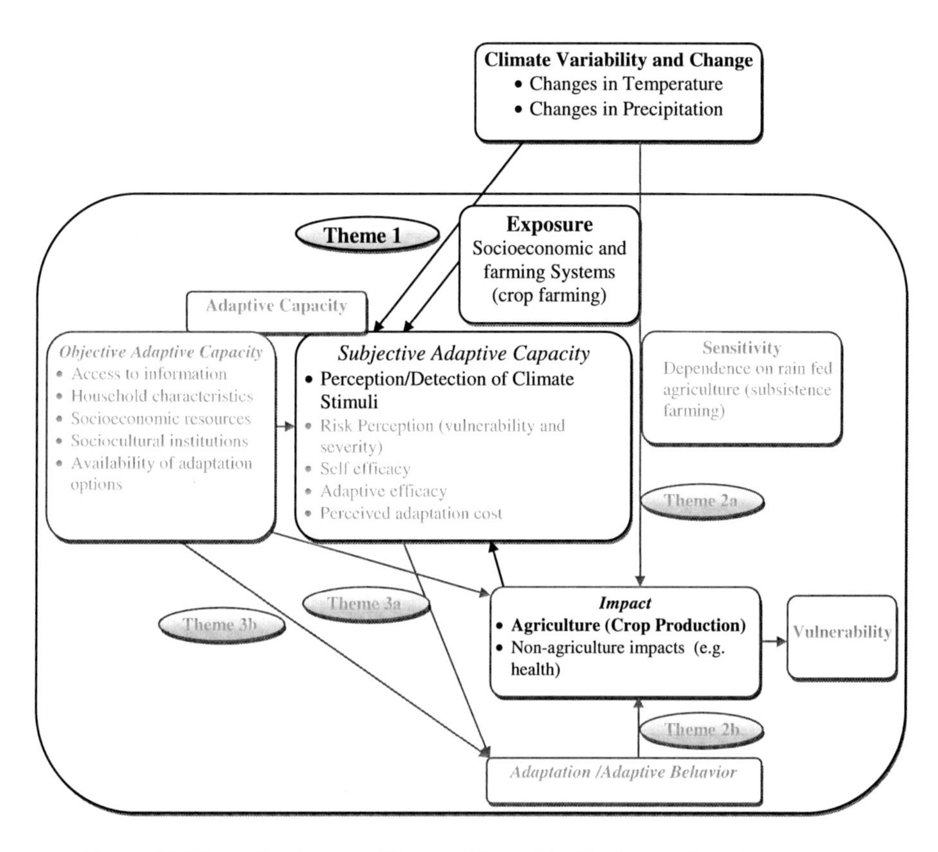

Figure 4.1. The major theme and issues addressed in this chapter (key elements are highlighted)

Scientific study of climate apparently involves some major steps that help to document relatively comprehensive knowledge on the subject matter. The first step involves detection of climate change and/or climate variability. The next step goes with attribution of the changes. The third step entails the study of impacts. The fourth step centers on response mechanisms. These steps were firmly employed in this study to dig out the overall understanding of climate variability and change among farmers.

While following these steps, both quantitative and qualitative data were used. As briefed in the previous chapter, the qualitative data comprises information obtained from focus group discussions and interviews. The quantitative data, on the other hand, encompasses data from the administrated questionnaire, meteorological recordings, and other secondary documents from Central Statistics Agency of Ethiopia (CSA).

While content analysis was employed in analyzing the qualitative data, descriptive and relevant statistical methods (regression, t-test and chi square) were used in analyzing the quantitative data.

In analyzing the qualitative data, recordings from focus group discussions and interviews were first transcribed verbatim. From the overall transcribed document and field notes, a list of texts and descriptions was prepared under the pertaining climate variables (namely rainfall and temperature), and the research's main topics of interest (detection, attributions and impacts). Next, texts and descriptions that have similar meanings and context under each main topic and climate variable were identified and coded manually. From this coding process, categories referring to similar behavior were formed and the interpretation was performed based on these categories. Whenever a detailed description was deemed necessary, relevant parts of the transcribed document were revisited as required. In the analysis, direct quotations, instances and photographs were also used to supplement or justify the interpretation.

Quantitative data, with aim of triangulating the subjective assessment of farmers, was analyzed descriptively through the application of relevant statistical tests. In analyzing meteorological data, a series of graphs (supplemented with regression functions) were constructed to depict the trend of climate variables (precipitation and temperature).

4.2. Demographic Characteristics of the Respondents (Sample Households)

Here only selected characteristics pertaining to heads of the households are presented before going to the main part of the analysis. Information on age and experience was in fact collected on continuous data form. But for the presentation here under, the information (data) was grouped into categories for the purpose of figurative depiction.

In rural areas of Ethiopia, male-headed households are dominant (CSA & ORC Macro, 2006). This is also reflected in the sample of this study. The overwhelming majority (about 84.4 percent) of the households are headed by males.

Table 4.1. Sex of heads of the households

Sex of the household's head	Frequency	Percent
Female	39	15.6
Male	211	84.4
Total	250	100

Source: Own field research data

The average age of the heads is 46 years and the minimum age is 20 and the maximum stands at 90 years of age. More than 68.4 percent of the heads are above the age of 41. In fact, the percentage increases to 84.8 percent when heads over the age of 35 are tabulated together. Having more farmers over the age of 40 years helps to exploit their knowledge built through the ages.

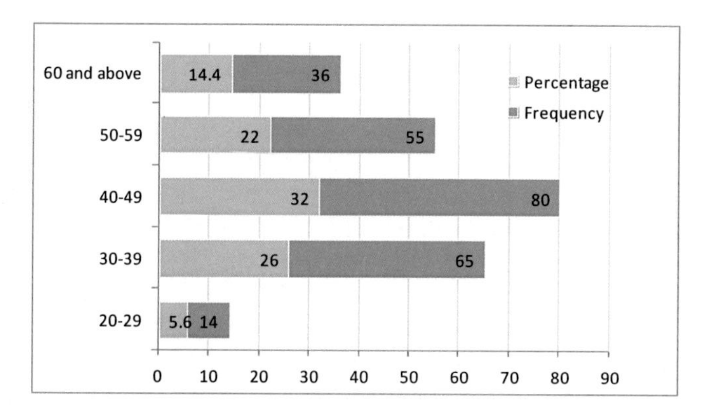

Figure 4.2. Age of heads of the households in categories

Source: Own field research data

As regard to education, as large as 62.8 percent of the heads had never attended school. Educated heads are a little bit more than half of the size of uneducated ones. Not more than 31.6 percent of the heads attended primary education (grade 1 to 8) and only 5.2 percent passed over to secondary level (grade 9 to 12). Although education levels are low, which is a typical characteristic of rural communities in developing countries like Ethiopia, most of the heads are rich in experience as measured by the number of years they spent on farming.

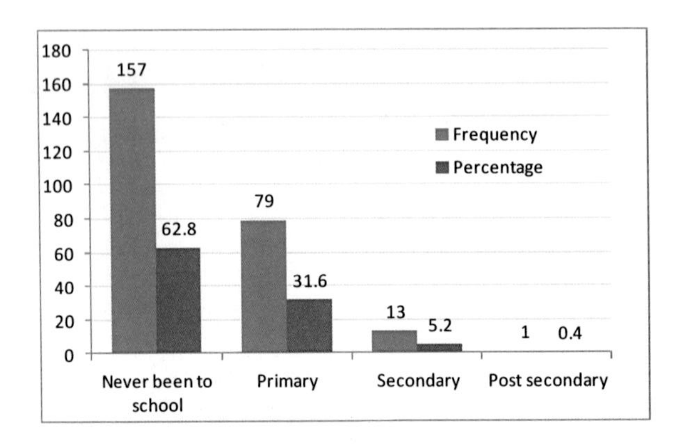

Figure 4.3. Education level of heads of the households

Source: Own field research data

Experience in farming goes up to 65 years with an overall average of 28 years. About 95.2 percent of the heads have at least 10 years of farming experience and 78.8 percent have at least twenty years of experience. Nearly 47 percent and 20 percent of the heads have more than 30 and 40 years of experience, respectively.

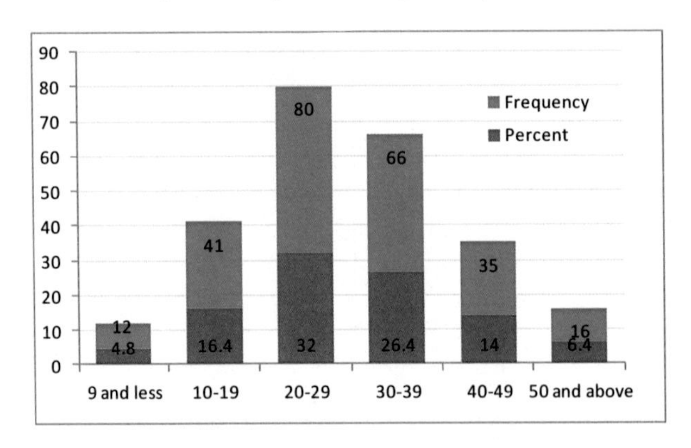

Figure 4.4. Farming experience of the household heads in years

Source: Own field research data

4.3. Farmers' Perception and the Trend of Climate Variables

This first part of the analysis examines the perception of heads of the households about the trend of precipitation and temperature in the last twenty years and compares perception with meteorological recordings in each of the weather stations. Twenty years was taken as a reference for two reasons. The first reason pertains to meteorological recording which is more limited as well as relatively good in the last two decades. The second and the most important reason goes with memories of farmers. Since farmers were asked about a perception that involves long years of insights, assisting them to retrieve their memories through some unique events or special occasions could be helpful from which time they would assess the trend of climate variables. In this case, the year 1991, a year in which change of government took place in the country, was taken as a reference from when on farmers assess the trend of precipitation and temperature.

A total of 250 heads, each representing a household, were interviewed. With the exception of one missing response, all households gave responses to the respective questions referring to the trend of precipitation and temperature. The first question posed was about as to how the amount of precipitation has been performing over the past twenty years. Accordingly, 96.4 percent of the households noticed changes in the trend of precipitation. Out of the total households interviewed, 82.8 percent of them perceived a decreasing trend on the amount of precipitation, and only 13.6 percent noted the contrary, i.e. an increasing trend (Fig. 4.5).

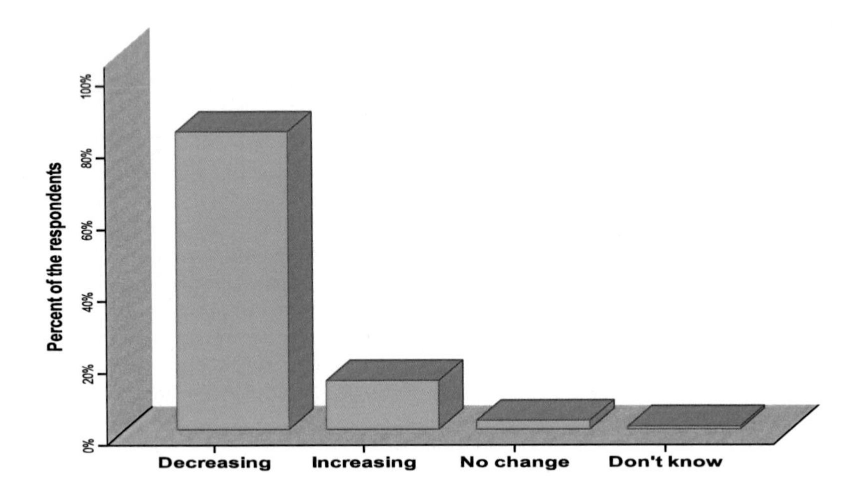

Figure 4.5. Perception of farming households about the trend of precipitation in the last twenty years

Source: Own field research data

At the outset, one can bear out the declining tendency of precipitation by looking to the sheer number of households who attested the declining trend of precipitation. However, unless the proportion of respondents who noticed the decreasing trend are statistically different from the other groups (those who noted the increasing trend or who noticed no change or cannot decide at all), it is hardly possible to come to conformity simply by the number of respondents. In addition, given the complexity of climate variability and change and the subjectivity of perception, the expectation in this study was nearly even proportion among the groups. As a result, a non-parametric Chi-square (χ^2)[46] test was administered to determine if there exists statistically significant difference between the groups of the respondents. The test statistics reveals that the proportions of the frequencies are significantly different to each other ($\chi^2 = 458.55$, $p < 0.05$ with df = 4). Further testing was also carried out to see if this significance difference stays consistent when less frequency groups ('no change' or 'don't know') were dropped

---

[46] Non-parametric chi-square is administered to compare observed and expected frequencies in each category to test whether categories contain the same proportion of values (whether there are differences in proportions among groups of a single variable or verify differences between observed and expected frequencies). The H_0 states that there is no significance difference between observed and expected frequencies or there is no significance difference between proportions of the frequencies (SPSS, 2007; http://academic.udayton.edu/gregelvers/psy216/SPSS/nominaldata.htm, accessed 11/07/2012).

and direct comparison was made between those who said 'increasing' and 'decreasing', as these two represent the largest groups. The output still confirms the existence of significant difference between the two groups ($\chi^2 = 124.19$, $p < 0.05$ with df =1).

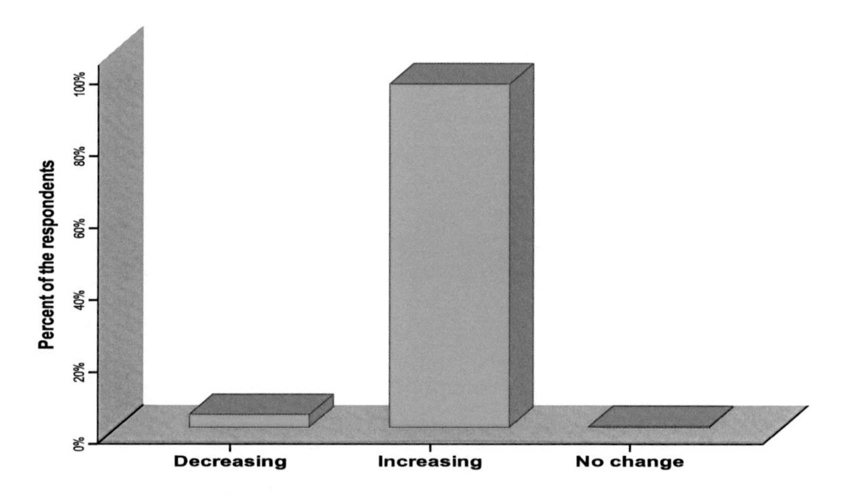

Figure 4.6. Perception of farming households about the trend of temperature in the last twenty years

Source: Own field research data

Similar question was posed to the households as regard to the trend of temperature, and a change was mentioned nearly by 99 percent of the respondents. As shown in Fig. 4.6., the great majority (95.6 percent) of the households perceived a rise in temperature and only 3.6 percent rather said the temperature has been getting cooler. 'No change' was mentioned only by 1 household (0.4 percent). Chi-square test also shows statistically significant difference between the proportions of the groups of respondents with $\chi^2 = 440.19$, $p < 0.05$ and 2 degrees of freedom, when all groups of respondents are considered together. Still the test turns significant ($\chi^2 = 213.31$, $p < 0.05$ with df = 1) when the two groups, namely who said 'decreasing' and 'increasing' were considered.

To make a conclusion about the trend of the climate variables based on perception (as one source of local knowledge), it would be sound first to verify perceptual judgment through available objective assessment method(s). Therefore, the accuracy of subjective assessment of the households was compared against long term (in this case 20 years) meteorological data of precipitation and temperature recordings of the nearby weather stations in the respective residence sites of households over the period of 1990 to 2009[47]. Graphic analysis of each of the five weather stations is presented in Figures 4.7 to 4.11. Regression and correlation were used to assess the trend of climate variables with time in years. The least square method was also used to quantify the trend of climate variables and the results are shown in Tables 4.2 and 4.3.

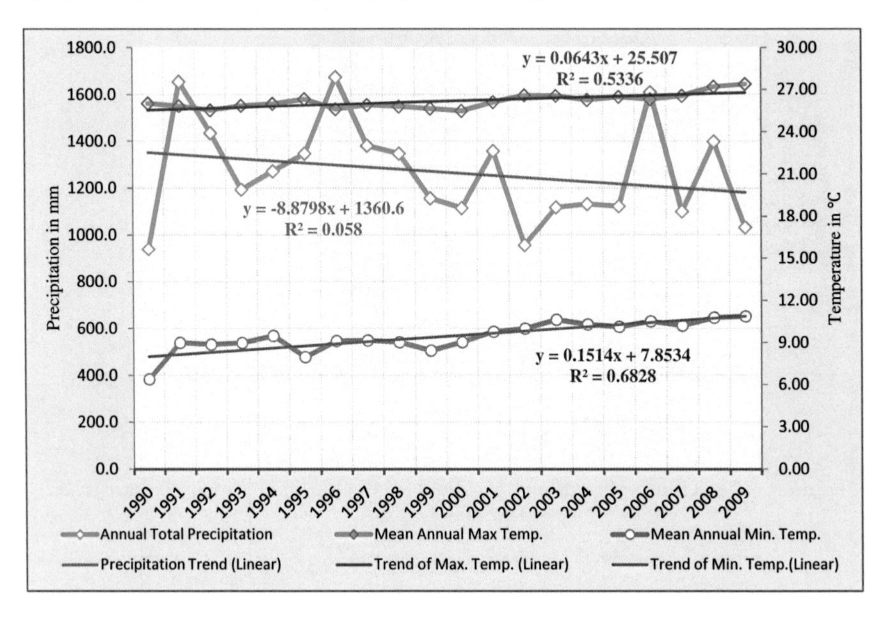

Figure 4.7. Trend of Precipitation and Temperature in Adet

Source: Constructed from a raw data of National Meteorological Agency of Ethiopia (NMA)

[47] Initially data from 1991 to 2010 was thought. But data for the year 2010 was not ready at the National Meteorological Agency at the time of data collection. Thus, for the analysis of meteorological recordings of two decades, data from 1990 to 2009 was considered. In the case of Arsi Negele, 12 years (from 1998 - 2009) were considered due to unavailability of data from July 1992 - December 1997.

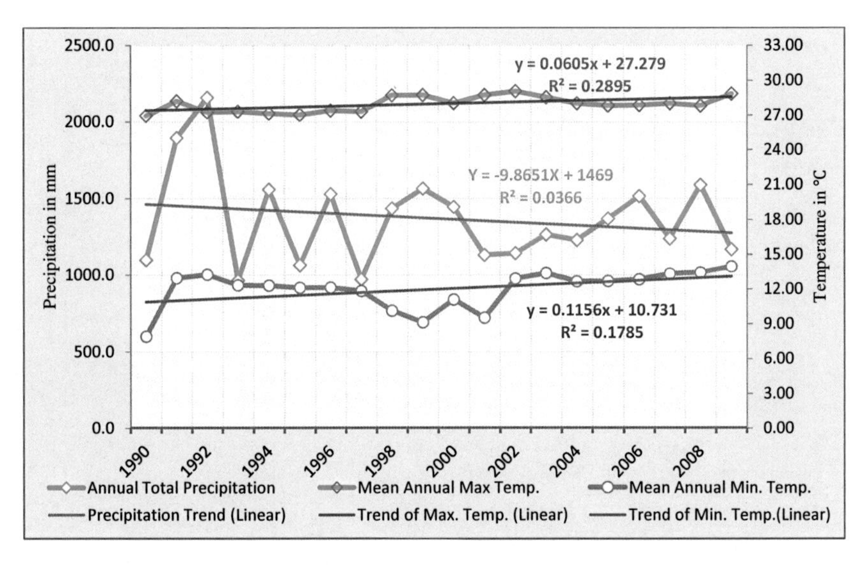

Figure 4.8. Trend of Precipitation and Temperature in Wereta
Source: Constructed from a raw data of NMA

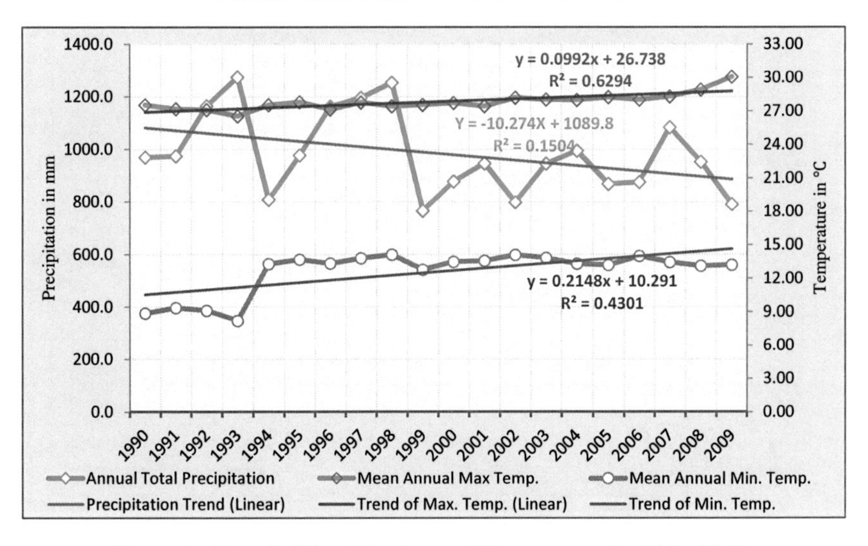

Figure 4.9. Trend of Precipitation and Temperature in Alaba Kulito
Source: Constructed from a raw data of NMA

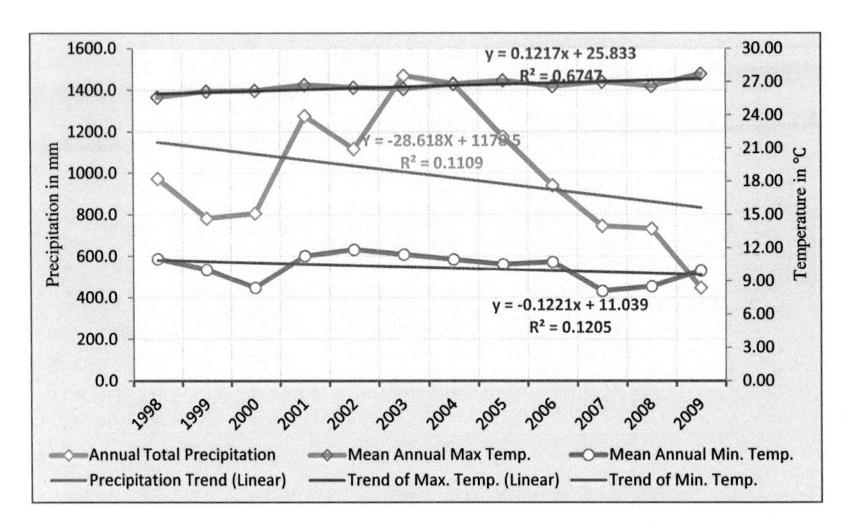

Figure 4.10. Trend of Precipitation and Temperature in Arsi Negele
Source: Constructed from a raw data of NMA

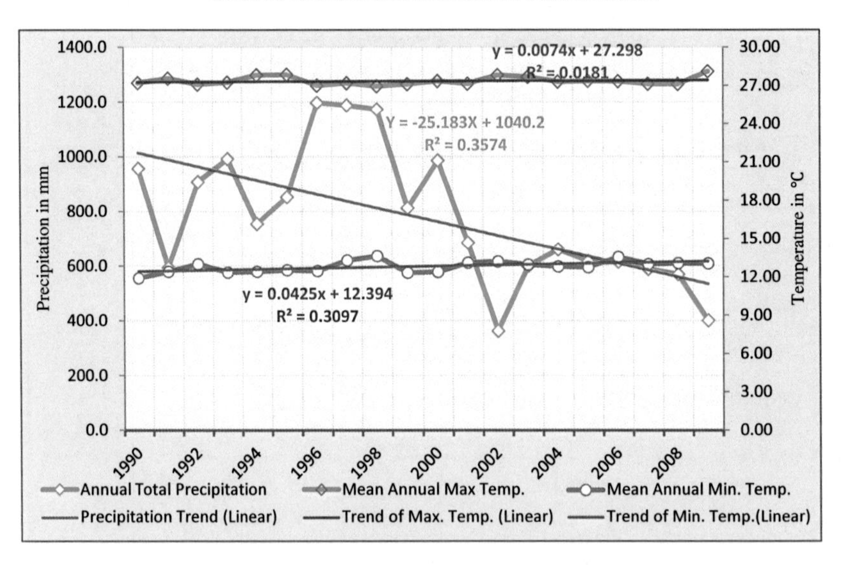

Figure 4.11. Trend of Precipitation and Temperature in Shashemene
Source: Constructed from a raw data of NMA

The regression analysis reveals that precipitation in all of the five weather stations is declining. Time (in years) explains 5.8 to 35.74 percent of the variation in precipitation. As can be seen in Table 4.2, precipitation decline varies from 8.88 to 28.62 mm per year, the lowest decline being in Adet and the highest in Arsi Negele stations. However, it is in Alaba Kulito and Shashemene stations that the trend is statistically significant at 10 percent and 1 percent levels, respectively. Similarly, correlation of precipitation with time is significant at these stations at 10 percent and 5 percent levels, respectively, with an overall negative relationship in all of the stations.

Assuming the negative sign in the trend of precipitation in all of the five weather stations, households' perception appears to be in accordance with meteorological recordings. Particularly, the statistical significance of the trend in Alaba Kulito and Shashemene weather stations strengthens the subjective assessment of the households. According to the information obtained from focus group discussions (discussed below in section 4.4), farmers in general agree with the decreasing trend of precipitation. But, they are also cautious of increasing occurrence of unexpected heavy rains that could mask the general declining trend of precipitation. Such occurrences in fact might influence the statistical analysis, as it (statistical analysis) lacks the capacity to capture the intensity and uniformity of rainfall over time. On the other hand, farmers assess precipitation not merely based on the amount but also consider its uniformity, intensity and pattern over time and space. Given the lack of capturing such important characteristics of rainfall in the statistical analysis, some weight has to be paid to the subjective assessment of farmers as the very clients of precipitation. Therefore, though the decline of precipitation in some of the stations turns insignificant, the negative sign in the change of precipitation over time could be a good indication that the households' perception about the trend is at least in line with the general trend of precipitation over time.

Table 4.2. Analysis of precipitation data from 1990 to 2009

	Precipitation in mm				
	Adet	**Wereta**	**Alaba Kulito**	**Arsi Negele**	**Shashemene**
Minimum	939.000	962.700	763.400	448.500	363.000
Maximum	1675.100	2158.500	1272.200	1470.200	1197.700
Mean	1267.345	1365.455	981.930	992.483	775.765
Std. Deviation	218.102	304.984	156.716	309.915	249.189
Correlation with time	-.241	-.191	-.388*	-.333	-.598**
Trend (per year)	-8.880	-9.865	-10.274*	-28.618	-25.183***

***, **, and * indicate 1, 5, and 10 percent significance levels, respectively.

Source: Estimation of a raw data obtained from NMA

Maximum and minimum temperature in all of the stations are visibly increasing as time goes on with the exception of minimum temperature at Arsi Negele station, where it shows a declining trend. Figures in Table 4.3 show that time and maximum temperature are positively correlated in all of the stations. All are significant at 1 percent level in Adet, Alaba Kulito and Arsi Negele stations and at 5 percent in Wereta station with the exception of Shashemene station where it is insignificant. Similarly, the rising trend of maximum temperature is statistically significant in four of the stations with three of them at 1 percent significance level. Here the exception is Shashemene station, where it is statistically insignificant. Over all, time explains 1.81 to 67.47 percent of the variation in maximum temperature of those stations where the trend is significant. The lowest variation is at Wereta station and the highest being at Arsi Negele station.

Table 4.3. Analysis of temperature data from 1990 to 2009

Maximum Temperature in °C					
	Adet	**Wereta**	**Alaba Kulito**	**Arsi Negele**	**Shashamene**
Minimum	25.490	26.910	26.500	25.610	26.960
Maximum	27.410	29.030	30.020	27.720	28.120
Mean	26.184	27.915	27.780	26.623	27.378
Std. Deviation	.520	.666	.740	.534	.328
Correlation with time	.732***	.538**	.793***	.822***	.133
Trend (per year)	0.064***	0.060**	0.099***	0.121***	0.007

Minimum Temperature in °C					
	Adet	**Wereta**	**Alaba Kulito**	**Arsi Negele**	**Shashamene**
Minimum	6.410	7.870	8.170	8.100	11.930
Maximum	10.890	13.910	14.080	11.860	13.660
Mean	9.444	11.944	12.546	10.243	12.840
Std. Deviation	1.085	1.619	1.936	1.269	.452
Correlation with time	.827***	.423*	.656***	-.348	.556**
Trend (per year)	0.151***	0.115*	0.214***	-0.122	0.042**

***, **, and * indicate 1, 5, and 10 percent significance levels, respectively.

Source: Estimation of a raw data obtained from NMA

As regard to minimum temperature, the rising trend is significant in four of the stations, namely Adet, Wereta, Alaba Kulito and Shashemene. At these stations, change in time explains 17.85 to 68.28 percent of the variation in minimum temperature. The exception here is Arsi Negele, where the trend goes declining but turning statistically insignificant. It is evident that the rising trends of both maximum and minimum temperature are significant at Adet, Wereta and Alaba Kulito stations. At Shashemene station, although maximum temperature is statistically insignificant, both minimum and maximum temperature are indeed rising. The exception is only Arsi Negele, where opposing trend is observed. At this station, minimum temperature has been declining insignificantly whereas maximum temperature has been rising significantly. Such significant rise of maximum temperature presumably masks or outweighs the insignificant trend of minimum temperature. Accordingly, the perception of farming households that temperature is rising coincides with temperature recordings of the weather stations.

4.4. Farmers' Description, Detection and Attribution of Climate Variability and Change

Households' perception about the trends of precipitation and temperature is one aspect to detect and document changes in the local climate. It also helps to demonstrate the insights of local knowledge in studying local climate. However, thorough understanding of local knowledge and its detection mechanisms should involve systematic exploration of farmers' thoughts and experiences beyond trend analysis. The discussion in the previous section lacks this aspect and without it, our understanding of the changes will only be shallow or might not be as such meaningful. This section, thus, builds on local detection of climate variability and change based on reflections obtained from focus group discussions and individual interviews.

At the outset, it has to be noted that farmers did not seek to differentiate between climate variability and change. Unlike the academic world which incessantly tries to put a demarcation line between the two terms, farmers (as noticed thoroughly both in the interview process and focus group discussions) appeared to have no interest to differentiate or have no awareness on the distinctions. As a result, in expressing the general condition of the changes and fluctuation in the climate, they interchangeably use the words *'ye ayir lewit* or *ye ayer mezabat'*, and *'ye ayer huneta melewawet* or *mekeyayer'*, which in the academic world the first referring to climate change and the latter to climate variability. Nevertheless, they clearly described the changes they have observed in the climate through its major elements, i.e., rainfall and temperature. As per the views of farmers, there have been pronounced fluctuations and changes (*'ye ayir lewit* or *ye ayer mezabat'*, and *'ye ayer huneta melewawet or mekeyayer'*) in the local climate as observed through its precipitation and temperature ingredients in the last several decades. While there were relatively steady reflections among farmers about temperature which had been consistently expressed by its increasing trends, conversely there were various characterizations of rainfall that were critically viewed both in focus group discussions and interview sessions. Each of the climate variables are discussed as follows.

Rainfall

As a major climate variable that has an upper hand in their livelihoods, farmers described rainfall performance in various forms of its manifestations comparing it with the old times. Old time here refers to at least 3 to 4 decades back, which farmers denote it as their childhood time and/or their fathers' time. In terms of information linked to it, the time frame also refers to the knowledge that they received from their forefathers and the past generation. They typically portrayed the features of rainfall in indicator words such as onset, cessation, duration, drought, flood, intensity, distribution, cloud, timing and volume. For analytical purpose, such expressions were condensed into four categories of precipitation's behavior and are presented as follows.

a. The timing of rainfall

Timing of rainfall refers to onset and cessation of rainfall, and unpredictability of its nature. According to farmers, onset and cessation dates of rainfall were relatively stable in the older times. These days, however, they tend to be more unpredictable, in some years being early and in some other years being late. They stated that from time to time, they continued to notice the occurrence of more fluctuations on the timing of rainfall. They further explained that seasons that receive rainfall in the right time of onset and cessation are dwindling these days as compared to the old times. They were more inclined to pronounce an unusual lateness on the start, and early withdrawal of rain.

b. The duration and pattern of rainfall

Duration of rainfall along with its amount represents the major area of concern that farmers attached to rainfall of recent (their) time. Characteristics such as interruption, lack of continuity, and shortness of rainfall at a time and season are the main reference terms that farmers used to describe the duration and pattern of rainfall. They said that as compared to the old times, the rain quite often lacks continuity and more likely interrupts at any point in a season. This was what they had said in a focus group discussion in BNB.

> *FGD in BNB: As we heard from our fathers, we learn that the rain was so generous in their time. It usually used to come on time and rarely interrupts in a season.*

For farmers, the rainfall is whimsical (unpredictable) that one tends to be cautious when dealing with it in general. In older times, the rain was said to stay as long as 6 months and sometimes even more than that. This has now diminished significantly up to 4 months and sometimes even less than that.

The farmers further stressed the changes they noticed with regard to the time span of rainfall in a day or a week. In their fathers' time and even in their childhood, as they expressed, the rain was quite a gentle one that sustains for longer hours. It was a kind of rainfall which they proclaimed very suitable to the soil. There were plenty of days that they never see the sun at all while it sheltered behind rain clouds. There were several weeks in a season, particularly in the months of August and July, when the rain was used to shower in each days of a week without interruption. Here is an account of a 42 old farmer in BNB who said, "*I remember, in old times we used to prepare and clear the land, while the whole day was cloudy and prone to rain*". But now, according to the farmers, the rain is intermittent and it tends to downpour in a short period of time and usually followed by interruptions for days, or weeks to the worst. After disappearing for days or weeks, it then comes and heavy rain (sometimes coupled with hail) pours down again. Mostly such intense rains are likely followed by strong sunshine.

c. The amount of rainfall

The amount of rainfall in terms of intensity is another major characteristic of rainfall that was critically viewed by farmers. In fact, the amount of rainfall captures not only intensity but also its trend over time and other characteristics which are directly related with amount such as flood and drought. Generally, farmers have developed the sense of viewing the rainfall being short in duration and getting more intense per time duration. There is also clear consensus among farmers about the overall diminishing tendency of rainfall in its amount of over time. Such perception replicates the perception of heads of the households as already discussed in the previous section. Here the attention is, however, on the other characteristics of rainfall that typify its amount. Other extreme features that are linked with amount are also discussed in separate part next to this briefing

As indicated earlier, for farmers, the rainfall of current era is more intense (defined by heaviness) than they had known it in their childhood. Here it has to be noted that farmers' expression of intense rainfall should be understood vis-à-vis the duration of the shower at a time. In this regard, the rainfall tends to be short in duration but intense in its strength. As a result, there is a tendency of more flooding now than the old times.

d. Occurrence of extreme events: drought and flood

Farmers maintained that they are aware of flooding even in the old times and also heard of incidences from their fathers. But, they marked out the flooding of recent years as more frequent and strong. Here is one example of a flooding incidence in one of the study areas that the farmers witnessed as a new event of flooding.

In a rural small town called, Senbete, which consists of about 800 households in Shala Wereda (RVLB), a flood occurred following a heavy rain early in the morning of April 22, 2010. It displaced 264 households. Of these households, 37 houses were completely destroyed. The flood washed away portions of roads, and wrecked four stores containing over 252 quintals of maize. It had forced the displaced people to stay 3 months on aid. As per the secretary of the Kebele Administration, he had never seen such flooding incidence in his lifetime and the elders also affirmed that they had never encountered such type in their lifetime. Similarly, a flood had occurred in 1988 in this same place but not as destructive as this one.

Figure 4.12. Some of the houses damaged by the flood of April 2010 in Senbete (Shala Wereda)

Source: Own photograph from the field research

Figure 4.13. A mass of land washed away by the flood of April 2010 in Senbete (Shala Wereda)

Source: Own photograph from the field research

It is paradoxical for farmers to face such destructive flood in an area where it was hit by drought just only two years preceding the flood. In fact, such incidences might not be surprising given the increasing trend of flood in Ethiopia in recent decades. Formerly, the country was well known of drought but not as such pronounced with widespread flooding. For instance, floods which destroyed property and claimed lives in 1997 in several parts of the country were considered to be the worst in 40 preceding years (Reuters, 12/5/1997 in Wolde-Georgis, 2000). Other recurrent floods had then followed in the North, South and Eastern parts of the country to the level that could be characterized as a national catastrophe (Ayalew, 2007). To cite some socioeconomic damages that are recognized nationally, it killed 498 people in 2006, and resulted in economic loss of $ 2.7 million in 1999, $ 6.2 million in 2005 and $ 3.2 million in 2006 (EM-DAT, http://www.emdat.be/result-country-profile, accessed 05/08/2012). So, the growing frequency of flood that farmers observed in the study areas could be viewed as part of emerging floods throughout the country. Similarly, farmers in BNB stressed of frequent diurnal and nocturnal floods (excessive runoff) that follow short-lived heavy downpours that usually tend to be sudden and unpredicted.

As regard to drought, there is also a general consensus among farmers that its frequency has been seemingly growing. But their expression is not in strict form of drought[48] (total failure of rainfall or deficiency of precipitation over prolonged period of time, which is meteorological drought). They expressed drought more of having agricultural drought, where a kind of drought weeks (intermittent droughts) sustain in between periods of rainfall events. There are no as such extensive droughts that cover the whole season except those very important occurrences at a time of delay in a season (early drought) and early cessation of rainfall (terminal drought), in addition to intermittent droughts. In the words of farmers, such occurrences of drought are getting worse from time to time and are the most serious impediments for crop production.

Temperature

Farmers described the changes of temperature consistently in terms of increment of air temperature, warmness of days and strength of the sun. In such characterization of temperature, a consensus among farmers was noted in the entire study villages. According to the farmers, temperature has gone through pronounced changes given the era of their childhood and their fathers' time. They indicated that after 4:00 - 5:00 o'clock Ethiopian time[49] in the morning (10:00 - 11:00 Central European Time), the day tends to be very hot, the air gets drier and the sun feels quite harsh. They described the strength of the sunshine as *'anat yemisentik tsehay'*, which is equivalent to say as ' the sun that harshly hits (emphasis on the heat that penetrates) the head'. So, for farmers such feeling from the sun has increased considerably in recent decades.

In a nutshell, it could be possible to infer that the local climate has changed (based on farmers' knowledge) in the ways discussed above. At this point, it would be logical (also to address our curiosity) to cross-examine this conception (farmers' account) with objective means (meteorological data) as done in section 4.3. However, lack of sophisticated long term daily meteorological data stands as a big impediment. Even, in the availability of data, some of farmers' descriptions might not be possibly captured through objective measurements. Therefore, other methods were used to verify farmers'

---

[48]There is no one agreed upon definition of drought as it depends on the form and meaning that people attach to it based on local climate conditions. Because of this, it is usually defined in three main ways.

Meteorological drought: is defined as an extended period involving below average precipitation.

Hydrological drought: refers to the low level of water (below average) in ground and storage supplies such as rivers, reservoirs, lakes, and aquifers. It might occur either due to low levels of rainfall or increased usage of water resources by humans.

Agricultural drought: is a drought that involves shortage of precipitation sufficient to adversely affect crop production or lack of enough moisture to maintain average crop production (ESPERE Climate Encyclopedia, English full version 2004 - 2006. http://espere.mpch-mainz.mpg.de/documents/pdf/Encyclopaediamaster.pdf; http://md.water.usgs.gov/drought/define.html, accessed 21/08/2012).

[49] In Ethiopia the day light has 12 hours and the night has also 12 hours. The day starts at 12:00 local Ethiopian time, which is 6:00 CET. The day ends at 12:00 local time, which is equivalent to 18:00 CET.

accounts. One method was to examine their detection experiences through observable changes or concrete examples. An easily accessible alternative was to resort to local experts in the field and re-examine farmers' knowledge through the views of these experts' (agricultural development agents). Since the former method involves exploration of real and vivid experiences encountered by farmers, it requires discussion at some length. Hence, it is treated at some length in the next subsection. With regard to the latter method, there was no account found among experts to refute farmers' characterization of the climate. Rather, experts' views were found to strengthen farmers' accounts. Alike farmers, experts clearly believed in the rise of temperature. With respect to rainfall, they characterized it being more variable from year to year, tending to be less predictable, heavy and sometimes accompanied by flood, inconsistent in its onset and cessation, sometimes inadequate for crop production, and sometimes coupled with interruptions at any point in the season (Figure 4.15 below). This depiction of rainfall is also found to be in line with farmers' description.

4.4.1. Detection through Farmers' Concrete Experiences

So far, farmers' detection knowledge was analyzed based on perception and climate description based on long-time observation. Perception might be apparently influenced by culture, among other factors. Observation could also be influenced probably by culture when describing and interpreting it. Both perception and observation might also be influenced by recent events in local climate. To minimize these drawbacks, one further step but simple technique was introduced to scrutinize farmers' detection knowledge. This step involved examination of concert experiences offered by farmers.

One simple technique to see the changes of temperature, as pronounced by farmers, is to examine the longevity of fermentation and preservation of local staple foods and beverages in older times and the present. This technique is vivid as it is very easy to assess, involves day-to-day activities and is meaningful for farmers. These advantages would lead us to document farmers' concrete experiences. The main disadvantage could be the bias that might arise due to the respondents' awareness of the interviewer's intention. As described in the methodological part of the previous chapter, women were interviewed in this regard tacitly without making them aware of climate variability or change as the study's intent. As a matter of fact, at some point two other women were interviewed while being aware of the subject under discussion, in case to see how that might have an influence in their assessment. But, no difference was observed with the attempt. In both cases, they reported marked changes currently in the length of food preservation and fermentation practices as compared to 2 - 3 decades ago. The following two tables (Table 4.4 and 4.5) show the reported changes in fermentation and preservation longevity of local foods and beverages in the respective regions of the study.

Table 4.4. Longevity of food and beverages' fermentation and preservation in Blue Nile Basin Villages

Food/ Beverage Item	Unit of Measurement	Time Frame	1	2	3	4	5	6	7	8	9	10	11	12
Tella[50] malt fermentation	Day	Old Time						*10 to 12 days*						
		Currently						*3 to 5 days*						
Longevity of Tella drink[51]	Day	Old Time						*4 to 5 days*						
		Currently						*2 to 3 days*						
Longevity of Tella malt[52]	Month	Old Time							*Up to 6 months*					
		Currently							*Up to 2 months*					
Injera[53] dough fermentation	Day	Old Time								*A maximum of 7 days*				
		Currently								*A maximum of 4 days*				
Longevity of baked injera	Day	Old Time								*A maximum of 7 days*				
		Currently								*A maximum of 3 days*				

Source: Own field research data

Table 4.5. Longevity of food and beverages' fermentation and preservation in Rift Valley Lakes Basin Villages

Food/ Beverage Item	Unit of Measurement	Time Frame	1	2	3	4	5	6	7	8	9	10	11	12
Keneto[54] malt fermentation	Day	Old Time								*3 to 7 days*				
		Currently								*1 to 3 days*				
Longevity of Keneto drink [55]	Day	Old Time									*6 to 8 days*			
		Currently									*2 to 4 days*			
Injera dough fermentation	Day	Old Time				*2 to 3 days*								
		Currently				*Within 1 day but also up to 2 days*								
Longevity of baked injera	Day	Old Time					*Up to 4 days*							
		Currently					*Up to 2 days*							
Longevity of baked bread	Day	Old Time										*Up to*	*10 days*	
		Currently								*Up to 7 days*				

Source: Own field research data

[50] Local beer in the Blue Nile basin villages.

[51] Finished (ready to drink) product.

[52] Already fermented Tella.

[53] Local flat bread which is a staple food in Ethiopia.

[54] Local beer in the Rift Valley Lakes villages.

[55] Finished (ready to drink) product.

As seen in the tables, there is a clear change in fermentation and preservation longevity of local foods and beverages. Such changes could be due to technological and/or climate change. One of the advantages of interviewing women without being aware of the researcher's intention lays in exploring potentially underlying causes (climatic or non-climatic) behind the changes. To identify this, the women were further asked to point out the reason (s) behind the changes. Their answers were unswervingly '*muketu new*', which means 'it is warmness of the air temperature'.

In a nutshell, farmers' observation shows that the climate has changed in the ways discussed above, and some of their detection experiences also strengthen their observation. It was also observed throughout the field work that there is firm belief among farmers about the change of climate in their respective localities. Such firm belief pushes further to raise one of the questions in the study of climate variability and change. To what factors do farmers attribute the changes already observed in the climate? The next subsection deals with this question.

4.4.2. Farmers' Attribution of Observed Changes in the Local Climate

So far, the discussion centered on detection experiences of farmers. If farmers believe in the occurrence of changes and unusual behavior in the local climate, they should have some underlying causes that they attach to the changes. Understanding the causes is essential in designing any form of response mechanisms to offset the impacts. Similarly, farmers take the root causes into consideration while attempting to implement adaptation strategies. For instance, if they believe that the cause is a supernatural force, this perception by itself might limit their counter actions, and thus they might resort to cultural and ritual practices that might have nothing to do with concrete adaptation actions. Hence, bringing forth the major reasons that farmers associate with the perceived and described changes in the climate is very important to deeply understand the process of adaptation in these communities.

According to farmers, the causes behind the changes are the following.

a. Population increment

As described by farmers, population in the respective Kebeles (villages) is highly saturated as compared to their fathers' time. Here is what a 46 old farmer from BNB had said, "*At that time (referring to the old time) people were few. Look, I myself became a head of five*". Similarly, an elderly farmer from RVLB briefed his story while linking population pressure with the changes he observed in the local climate.

> *I have twenty children. I do not possess as many cattle as my grandfather to help raise my children. I have already sold out my cattle for the sake of raising the children. Either, my plot of land could not be a much to them.*

A historical account by an elderly farmer, aged 76 from RVLB, clearly connects how population pressure acts as a source in aggravating the changes in the local climate.

> *In the old time, there were forests within the distance of 1 to 2 kilometers from our village. The air was fresh. There was also good rain at that time. Now, those trees and the forest are already cleared. We, ourselves, cleared the forest foremost because we became many.*

The farmers clearly stated that due to burgeoning population, communal grazing sites, tree and grass lands, swampy plains, moisture places and marginal lands by now are either settled or used for agriculture or degraded due to over grazing and over utilization. The rest of swampy and moisture areas are now either dried or filled with sand. These areas, on the accounts of farmers, were vital in maintaining the balance of air temperature of the surrounding and the environment in general. They further stressed that due to the destruction of these vital communal lands, the balance of nature has been disturbed, and exposed them to distressing happenings already observed in the climate.

Population growth and expansion of agricultural land in Ethiopia clearly supports the account of farmers. The 2007 Population and Housing Census of the country put the national population growth at an average annual rate of 2.6 percent between 1994 and 2007 (CSA, 2008). This growth rate marks Ethiopia being among the nations with bourgeoning population in the world. Given the great majority of the population living in rural areas (83.9 percent) and its growth proportion (2.26 percent), demand for new land openings to meet food requirements would be immense. The figure below, Figure 4.14, shows how cultivated land area in the country has been increasing in the last 17 years among private peasant holdings in rural sedentary areas. Land coverage of grains[56] in 2010/11 has increased by 69.86 percent from that of 1996/97. In these years, grain coverage of land has expanded averagely 355, 600 hectares per year. Where could such huge hectares of land per year in rural areas possibly come from? No doubt that it would be from clearing and reclaiming tree and grass lands, grazing sites, moisture places and marginal lands. This is what farmers interviewed in this study attempted to connect growing population with the pressure it creates on land and its resources, which in turn plays an important role in affecting local climate.

[56] Grain constitutes cereals, pulses and oilseeds. Grain by far is the major source food and income at household level. It also contributes to the country's foreign currency earnings. It comprises the higher share of the country's total crop production. For instance, in 2010/11 production year, it accounted about 89.10 % of the total crop production (CSA, 2011).

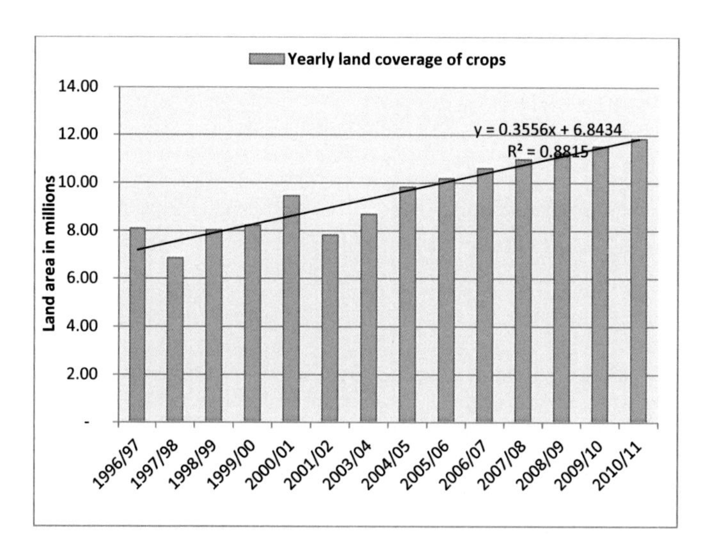

Figure 4.14. Land coverage by crops

Source: Compiled from Annual Agricultural Sample Survey of CSA

b. Deforestation

The above quotation from a 76 old farmer from RVLB could also serve to assert how farmers regard deforestation as a cause to the changes observed in the local climate. In older times, as reported by farmers, hills, stream sides, river banks, and several fields were covered by trees and there were considerable areas of forest and bush land. Due to increased demand for more land and firewood[57], trees and forests (which are instrumental in attracting moisture and stabilize the temperature, as described by farmers themselves) had vanished. In connection to this, a farmer aged 43 from BNB said the following:

> *I am now 43. I no more see those trees which I used to see when I was child. Even I do not find the trees I used to see just as near as 15 years ago. Those trees are not in place now. They are not found even in the gorges and the fields. We cut them. It is due to such act of ours that the climate has been changing, and the temperature has been rising.*

Lack of protection to the environment (inability to plant trees as replacements for the cuttings and taking care of the existing ones) was also mentioned as a contributing

[57] Firewood is the sole source of heating and cooking in rural Ethiopia.

factor along with deforestation. In a focus group discussion, a farmer plainly stated as: "*I could say that it is we who are responsible for the changes that occurred to the climate. It is we that contributed for the rise of the temperature*".

c. God's act

Although farmers blame their actions as a root cause for changes in the climate, still they maintain God's will taking the upper hand. This articulation was observed in all of the focus group discussions with the exception of one focus group discussion in BNB, where farmers' own actions (deforestation and population pressure) were predominantly blamed. Interestingly in that focus group discussion, the participants were found to be relatively literate ones (at least read and write) and composed of the younger generation. In another aspect, there was also a tendency among farmers to attribute extreme cases such as drought, hailstones and flood to God, and other gradual manifestations (a good example is hotness of days, lack of moisture in the air) largely to human action (in this case their own).

Further question was posed to farmers to understand as to how they associate the changes with God. Their views could be classified into two classes of thought and could be termed in this study as: the *notion of punishment* and *lack of stewardship*. As per their explanation, when people fail to obey God's words and commands, He punishes them through drought, flood, thunderstorms, hail, strong winds and plague of worms. The *notion of lack of stewardship*, on the other hand, links the changes with people's failure to protect the nature that God has given to them. There is a sentiment among farmers that people these days are becoming more egocentric and demanding, and living is also becoming more expensive. Consequently, everybody tries to make the most out of nature to fulfill his/her demands while paying little or no due regard to nature. Such breach of human's stewardship of nature has led them to the changes that they have faced now. The notion of stewardship, in fact, is a cultural explanation of human induced causes described above, such as deforestation.

Generally, from the discussion so far, it is evident that farmers relate the changes in the climate with human activities (deforestation and population growth) and cultural belief (God). While emphasizing on the Devine power, the cultural belief also relates the changes to human activity (in the form of lack of stewardship), which still shows the importance of anthropogenic causes. But, does this attribution of farmers relate with that of experts?

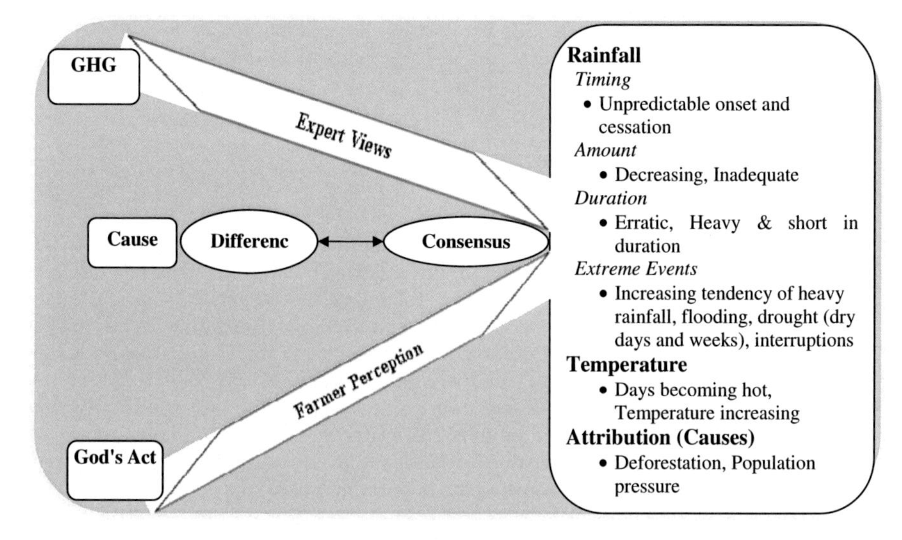

Figure 4.15. Farmers' and experts' description and attribution of changes in local climate

Source: Own construction from the field research data

As depicted in Figure 4.15 above, both farmers and experts (local experts mentioned in chapter 3) share similar views in characterizing the rainfall and temperature. However, in attributing the changes, they entertain both similarities and differences. They have similar ideas in the sense that they both believe in anthropogenic causes. But the scope of anthropogenic cause is limited to local actions when it comes to farmers and goes beyond that for experts. While maintaining the role of local human actions, experts prominently link it with greenhouse gas emissions. Major difference among farmers and experts, however, is marked in the cultural aspect of the cause. For farmers, God has place in the materialization of the change, the point where experts depart from farmers. Such difference is not surprising given experts' education level and proximity to the climate change discourse and its information outlet, and conversely the role of culture among farmers and their detachment from the discourse.

As stated repeatedly, despite the potential influence of culture, farmers have profound environmental knowledge that they developed over time through direct experience and observation. They have also a bulk of accumulated knowledge transferred to them through generations. Based on this experiential knowledge, farmers do come to in conformity with the scientific discourse that asserts the changes as more of anthropogenic. In fact, farmers do carry on the belief of God's will as well. Still in this line of belief, they emphasize lack of stewardship as a cause, which strengthens the human role in the observed changes.

4.5. Farmers' Description of Impacts of Climate Variability and Change

It is elaborated earlier that farmers described the changes they observed in the climate through its major elements, i.e., rainfall and temperature. It is also noted that although they clearly described the changes through the specific climate elements, they did not seem to differentiate literally when it comes to the terms, climate variability and climate change. Nevertheless, some differentiation and meaning of these two terms comes when one looks thoroughly into their explanation about the impacts. Rather than the literal meanings embedded in each of the climate manifestations, one could notice practical conceptualization and differentiation of these terms through specific climate causes and strength of impacts that farmers link to the type of climate's manifestations. This instance could take us to infer that farmers only make meaning of climate change or climate variability principally through the impacts that they face in their livelihoods. Their description of the impacts helps one to take a note to what form of climate events or manifestations (climate variability or climate change) that farmers link the impacts and which forms are more serious to them.

Having said this, it is possible at this stage to infer roughly that the state of climate has undergone through some changes as confirmed by farmers and complemented by meteorological data and experts. If this is the case, the next step would thus be assessing whether these changes have impacts on farmers' livelihoods. This could be addressed through qualitative approach by engaging farmers themselves and/or quantitative approach based on socioeconomic data. Qualitative assessment in particular helps to explore the existing impacts in a broad range of socioeconomic and ecological systems. On the other hand, quantitative approach, with the power of assessing the current impacts, is a convenient tool to predict future impacts as well. Since one of the interests of this study is to measure the economic impact of climate change on crop production, as the single source of farmers' livelihoods, it will address it quantitatively in the next chapter. But, before passing to the quantitative analysis, impacts from farmers' perspectives will be assessed first.

Farmers indicated that they experience repercussions from fluctuations and changes in the climate. They are increasingly concerned with impacts that are related to availability of water, agricultural production and health. In addition to these major problems, they also mentioned additional problems that seem to threaten their psychosocial establishments. Here under is a brief presentation of these problems.

a. Impact on Availability of Water

According to farmers, as rainfall and temperature have been decreasing and increasing, respectively, streams and rivers that used to flow the whole year are gradually lessening in their water volumes. Good examples offered by farmers include a river and a stream in Yilmana Densa Wereda of BNB. In fact, farmers in other study sites also believed that water flow in rivers and streams has been receding slowly over the years. Even in Shalla Wereda, some villages are forced to travel several hours (as much as 2 to 3 hours) to get drinking water due to water stress in the areas.

In Yilmana Densa Wereda, a river called Shina in Adet Zuria Kebele and Divdiv Stream in Mosobe Kebele, are said nearly to dry after January until the next rain season comes again on or after May. But formerly both of these water bodies were said to flow the whole year. Because of this, they stated that water for domestic usage and irrigation has been getting scarce from time to time. With respect to this, a farmer in a focus group discussion in BNB gave the following account:

> *For instance, until 1995 I was using irrigation from a small stream in my village. But today, let alone irrigation, drinking water has become scarce. Formerly, there were water bodies in the gorges but no more these days. At present, we struggle to protect the remaining drinking water bodies by keeping a fence around them. However, we feel that our protection now is like "a dog barking after the hyena has gotten away"*[58].

However, it became hard solely to link these accounts with climatic variability and change given an increasing practice of irrigation throughout the country to produce mainly vegetables and fruits. As part of the government policy to enhance food self sufficiency, farmers are encouraged to use rivers and streams for small-scale irrigation schemes. The farmers involved in this study also affirmed that small scale irrigation schemes are increasing alongside rivers and streams. Some rivers and streams are said to be used excessively by upstream farmers, and others being utilized beyond the carrying capacity so that valleys and gorges at downstream end up dewatered. Even though climate change and climate variability have clear impact on water availability, in this case, one finds very difficult to find the connection beyond doubt given the recent expansion of irrigation.

b. Impact on Health

One of the new phenomena that emerged in recent decades in one of the study areas is Malaria. Malaria, largely influenced by temperature, rainfall and humidity, is well known disease in Ethiopia which is predominantly associated in areas less than 2,000

[58] An Ethiopian proverb.

meters above sea level (MOH, 2008). But, in recent years, the impermeability of the highlands, which were considered to be free of the disease due to their altitude, seems to be breaking. An important example of farmers' account in Adet Zuria and Mosebe Kebele's of Yilmana Densa Wereda represents the incursion of malaria into highlands. Until the 1980s, malaria was not known to these areas, being highlands with an altitude over 2000 meters above sea level. Till that period, malaria was a rarely known disease under the name 'nidad'[59], before it was replaced by its nationwide name 'weba'[60]. So farmers recall the past as a free-malaria period and unlike today symptoms of the disease were barely known. They asserted that someone had the chance of contracting the disease only when he/she had visited malaria prone areas or Kolla (lowlands).

Here is an experience of a farmer in Adit Zuria Kebele, aged 46.

> *In 1978, I was seriously sick. There was no clinic in this area at that time. So, no one was having the knowledge of my sickness, as no one knew clearly the symptoms. While I was seriously sick, luckily a relative from a far area came to visit us, and the first thing he asked after looking to me was whether I had been somewhere else. They (his family) told him that I was in Bahir Dar[61] sometime ago. Then he told them that the disease was nidad, and it was after this that I went for a treatment.*

In these Kebeles, at the mid and end of 1990s, several people were reported to have been killed by malaria outbreak. Farmers, for instance, witnessed that they had counted 141 cases of deaths in 1999 in Adet Zuria Kebele alone. They further stressed that, at that time they had buried up to nine people per day. They recall that period as a malaria epidemic ever to hit their place. They acknowledged that had it not been the assistance of the government and other interventions, there could have been a disaster worse than they had encountered. Although, these days there is an aggressive malaria prevention program by the government, NGOs and the population at large, and consequently rare deaths cases, malaria has got a prominent place as one of the top diseases in the area. For instance, the available 3 years disease recordings at Yilmana Densa Wereda Health Bureau shows malaria ranking first among the list of 20 top diseases in the years 1999, 2000 and 2002. Similarly, the household data collected through questionnaire for this study shows that averagely 4 in the average household size of 6 in Yilmana Densa have a history of malaria infection with only 4 percent of the households (as seen in Figure 4.16) reporting no history of infection among the household members. Surprisingly, nearly two decades ago this area was believed to

[59] The name used to refer malaria in older times, when malaria was regarded as a lowland disease. Presently, that name is replaced by 'weba', with the influence of nationwide media and public health interventions, and clinical references to malaria.

[60] Weba stands for malaria in Amharic language.

[61] Capital of Amhara National Regional State, and is a malarious area.

have been malaria free, and now about 95 percent of the households have at least one family member with a history of malaria infection.

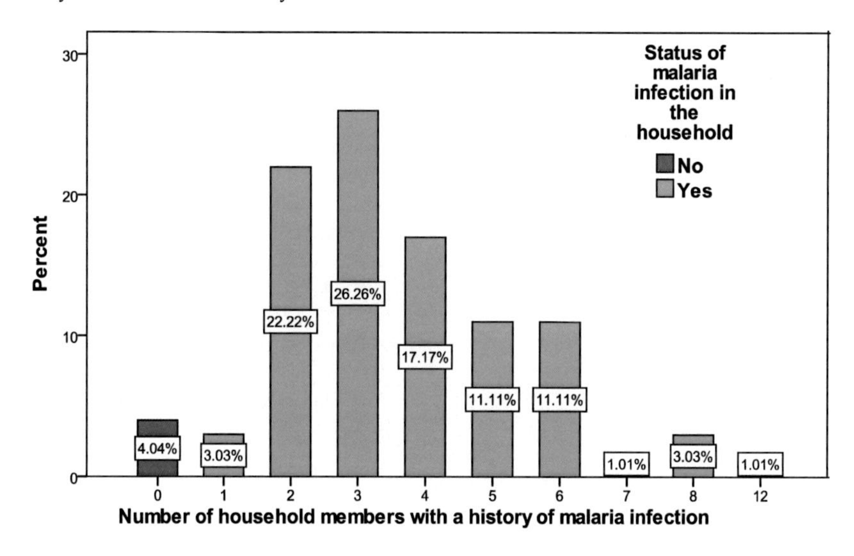

Figure 4.16. Household members infected by malaria in percentage in Yilmana Densa (Adet Zuria and Mosebo Kebele)

Source: Own field research data

The reason for the incursion of malaria to these areas exclusively rests on temperature increment, according to both farmers and experts. As a matter of fact, temperature increment, as evidenced by meteorological recordings presented at the beginning of this chapter, supports the account of both farmers and experts. Since meteorological recordings began in Yilmana Densa (Adet weather station) in October 1986, it could not be possible to see the changes in the temperature in the 1980s so as to compare with the period in which malaria was said to have begun occurring. However, given the relationship between temperature and malaria, the rise of temperature certainly takes the blame for the incursion of the disease in the area.

In the other four Weredas (Fogera, Shashemene, Arsi Negele and Shala), although malaria is not a new disease, farmers pinpointed seasonal climate having an upper hand influence over spread of the disease. In these areas, where malaria favoring temperature and humidity are already in place, rainfall importantly triggers malaria transmission. As per the farmers account, when rainfall is relatively good and has no interruptions, there is less likely of malaria infection. However, when the rain tends to be on and off, there is likely an increasing incidence of malaria. This explanation of farmers is also well supported by the head health officer at Shala Wereda Health Bureau, who attested the critical role of rainfall in malaria transmission in the Wereda.

Malaria as a disease has several implications. Obviously, it affects the health status of farmers, which in turn adversely affects labor productivity, where working time is diverted to recovery from sickness and care giving to the sick. In addition, there is a probability that money which could have been allotted to the purchase of production inputs is diverted to treatment of the sick and to buy insecticide-treated nets (ITNs). In fact, there is free medication and preventive program by the government, but farmers are still forced to buy additional ITNs for family members. They also incur transportation costs to get to health centers and other associated expenditure, such as for buying nutritious food for the sick and supplementary drug prescriptions. For instance, the household data of this study shows that averagely a member in a household gets sick of malaria 3 times per year and takes 10 recovery days per sickness and might spend 224 Ethiopian Birr per sickness for medication and related staffs.

c. Impact on Crop Production /Farming

Climate variability and change is said to have general impact on farmers at large and pose additional impacts on specific section of the farming communities. Farmers were observed vastly worried with the effects of climate variability and change on farming in general and crop production in particular. Since crop production is the principal means of livelihoods and the single source of food, the behavior of rainfall is a special concern to them. As a matter of fact, livestock production is also affected. However, based on their experience and understanding of local climate, farmers emphasized to a great extent on the impacts of crop production.

The rainfall behavior illustrated in Figure 4.15 has a direct effect on the growth of crop plants and eventually on productivity. Even if farmers recognized the general decreasing trend of rainfall (as discussed in section 4.3), surprisingly they appeared little worried of its impact when it comes to crop production. Their grave concern is not quite the amount (gradual decrement), but the timing and irregularity behavior of the rainfall. As outlined in the conceptual framework of this study (section 3.1, chapter 3), these characteristics of rainfall represent climate variability. From this perspective, farmers' grave concern is inclined towards the features of rainfall that characterize climate variability. This had been also noted throughout the field study where farmers noticeably link serious impacts of crop production with various features that characterize climate variability.

Irregularity in the timing (late onset and early cessation) and distribution over time (mainly interruptions at the middle) are said to have significant and sometimes unbearable effects on crop production. They stressed that such impacts of rainfall are affecting production from year to year. Although farmers affirmed the nature of rainfall as the most single factor affecting crop production, they are also recognizant of the role of temperature. For them, temperature importantly intensifies the effects of rainfall. As per their explanation, when there are prolonged interruptions, growing crops wilt not merely because of lack of rainfall but also of the hotness that bakes the soil (due to loss of soil moisture). Hence, for the farmers, rainfall is a single factor that affects farming, while temperature largely playing an exacerbating role.

As regard to specific impacts on a section of the community, climate variability and change is said to have extra impacts on female-headed households. Some female-headed households lack male labor for traction of land, sowing and harvesting. Since cultural and traditional norms in Ethiopia tacitly forbid women from ploughing with oxen, they are forced to rent out their land in the form of sharecropping to male farmers. When the timing and distribution of the rain tends to be inconsistent, these households are said to be badly hit by the effects. It was described that at the time of irregularities of rainfall (mainly on onset and cessation), rented plots of lands are more likely undermined in preparation and planting stages. Tenants (farmers who rented female-headed households' land) prioritize their own land in allocating labor and preparing the land as they would be forced to compete with the behavior of rainfall. It is later that they come to attend the land that they rented from female-headed households. In such cases, the rented land might lack timely cultivation and farm management decisions. Because of this, production in such lands are said to suffer the most.

d. Psychosocial Impact

Psychosocial impacts here cover feelings of confusion and bedevilment in social and cultural establishments. Due to unpredictability and fluctuations in the onset of rainfall, farmers especially get confused on the timing of planting. Even though, they usually depend on their senses and experiences, they are more uncertain of when to sow crops. Some farmers might plant earlier in the first weeks after the onset of rainfall and others wait and opt to evaluate the performance of the rainfall. This has implications on the productivity of crops. Depending on the performance of rainfall, there is time when early planters become more beneficiary and other times late planters taking the advantage, and vice versa. Such experiences put them sometimes in a kind of indecision and to face uncertainty in the dates of planting.

The other aspect described by farmers is related to tearing the social fabrics of helping each other. There has been a long tradition among communities to help each other in land preparation, ploughing, sowing and reaping. In these activities, priority has been given to disadvantaged groups (widows, disabled, old people without a working family member) where members of a community and relatives come to work for these groups of people in each farming season. In the current era, as farmers are in predicament with the behavior of rainfall, the disadvantaged groups are less likely to enjoy the statuesque of the social assistance.

Even if they get, the assistance comes late after farmers have attended their own farms. This has a negative effect on crop productivity of the disadvantaged groups especially if they fail to get timely assistance in the critical periods of crop planting. Here is an account put by an elderly man:

> *In our tradition unless we first prepare and attend the land of widows, disabled and the old, we fear God that He will halt the rain at all. These days, there is less regard to this custom as farmers are forced to compete with the behavior of the rain.*

Other accounts by farmers seem to complement this but also view the problem partly as the manifestation of an ever increasing expensiveness of living in general.

At first place, these groups of the farming community are socially disadvantaged who have limited or no access to productive labor, among other things, which in turn jeopardizes food security of the group. Hence, they are forced to fall back on community's assistance or in broader terms, social capital. Social capital[62] is quite extensively regarded as an important instrument in facilitating collective action to respond to disaster and development challenges (Dynes, 2002; Pelling, 2011). Particularly, social capital in the form of norms is considered as a powerful means to subjugate self-interest to the needs of the community (Dynes, 2002).

Similarly, in the climate change literature, social capital is regarded as a social apparatus to mitigate impacts and facilitate individual or collective adaptation to climate variability and change, and extreme events (Adger, 2001; Adger, 2003; Pelling, 2011). Such conceptualization has led the scientific community to gear towards understanding how social capital operates in dealing with climate variability and change just overlooking the reverse process. Nevertheless, social capital itself could be broken when it faces immense pressure from climate risks. Farmers' experience in this study typically attests this situation when survival is at a grave risk, social capital might not stand to its expectations. On the other hand, this situation could also be viewed as a ground to assert how climate variability and change exacerbates social vulnerability. Due to inherited social vulnerability (lack of family labor or economic capacity to hire labor), the disadvantaged groups have been living in a more susceptible condition. In such state of susceptibility, emergence of climate risks apparently sinks them to a

[62] Although it is contentious and multidimensional concept, social capital is widely understood as the social trust, norms and networks that facilitate collective action against disastrous or risk situation or solve development challenges (Woolcock & Narayan, 2000; Dynes, 2002; Pelling, 2011).

worsen situation of social vulnerability by dismantling communal values of looking after the disadvantaged groups of a community.

e. Impacts on Infestation of Insects and Weed

In addition to the aforementioned impacts, farmers also described the effects through infestation of insects and weeds. They said that as compared to the old days, insects invading crops have increased following the rise of temperature which created a favorable condition for the breeding and infestation of insects. Farmers said that without the use of chemicals, it is almost impossible to store grain at present.

In regard to weed, there is a general tendency among farmers to attest its overall spread as compared to the olds times. However, its infestation in the farm lands has decreased enormously. Since there is abundant labor per land size, labor is well allocated towards repeated ploughing and better farm management techniques, which consequently contributed to reductions in farm weeds. As per the account of farmers, their forefathers used to possess several hectares of land. However, due to population growth, the size has now considerably diminished even to a size of half a hectare per household. Therefore, following the abundance of labor per hectare of land, farmers today exert a lot of labor on managing the available small plots of land. Despite this labor intensive farming which contributed for the reduction of weeds in the farm lands, they confirmed that weeds that were typical to lowlands are already observed in their areas. Particularly, such behavior was pronounced in Yilmana Densa Wereda.

Table 4.6. Summary of impacts of climate variability and change as experienced by farmers

Impacts	Features of the impact
Impact on availability of water	- Gradual reduction of water flow in streams, rivers and gorges - Stream and rivers that used to flow the whole year tending to dry at the middle of the year - Scarcity of drinking water - Aggravation of water stress in some of the research sites
Impact on health	- New malaria invasion of the highlands - Increased infestation of malaria in malarious areas
Impact on crop farming	- Negatively affecting crop farms in general - Extra impacts on specific section of the community who rent their farmlands (e.g. female headed-households with no active labor force)
Psychosocial impact	- Confusion over planting dates - Breaking down social networks of helping the disadvantaged (widows, disabled and old people without a working family member)
Impacts on infestation of insects and weed	- Insects increasingly invading crops, particularly in storages - New weeds (typical to the lowlands) growing in the highlands

Source: Own field research data

4.6. Climate Stress Factors in the Face of Non-Climate Stressors

The analysis in the previous section revealed how climate variability and change has been affecting farmers. While farmers described the impacts based on their observations and experience, one can also argue that the impacts might be disguised reflections of non-climate factors that existed in the socioeconomic systems. As a matter of fact, one of the contentious and problematic issues in the study of impacts of climate variability and change is this issue where disentangling the impacts from that of non-climate factors goes beyond the limit of the existing knowledge. Nevertheless, based on farmers' own experience one could at least portray the tendency of impacts, i.e., which factors are more serious, if not visibly show the impacts of climate variables by disentangling from that of non-climate ones.

Farmers in developing countries obviously face a series of non-climatic factors that impede their livelihoods. Additional imposition by climate leads them to double exposure[63] (O'Brien & Leichenko, 2000). In such exposure, it would be very important for policy to identify exposure factors and prioritize them according to their importance. Being the victims of double exposure, farmers could depict the level or seriousness of the impacts from climate and non-climate factors. In this regard, farmers own experiences, as indicated above, could be referred to show the place of climate factors in the face of socioeconomic, institutional and technological stressors inherited in their social and economic systems. By doing so, it is possible to weigh the tendency of the impacts, as well.

Towards indentifying the tendency of the impacts and prioritize the stress factors, an attempt was made to see the position of climate factors as sources of concern in the face of non-climatic stress factors. To these end, a list of twenty climate and non-climate stress factors was developed through consultation of the literature and experts in the field. Farming households were then asked to rate the factors in a five grade Likert scale (Not a source of concern at all, 1; Limited source of concern, 2; Moderately a source of concern, 3; Highly a source of concern, 4; and Very highly a source of concern, 5).

The rating was finally analyzed with mean values of the state of concern attached to each stress factor. The descriptive results in Table 4.7 below show that among the top five stress factors (in order of their ranking importance), three of them belong to climatic factors and the rest of the two to non-climatic factors. If the first seven or ten factors are considered still the majority (five or six stress factors, respectively) belong to climate factors.

[63] Double exposure refers to confrontation of certain regions, sectors, ecosystems and social groups both with the impacts of climate change, and the consequences of globalization (O'Brien & Leichenko, 2000). Here in this study, double exposure is used similarly but conceptualized to consist of inherited socioeconomic, institutional and technological limits in addition to the consequences of globalization.

Table 4.7. The Mean value of stress factors in descending order

No.	Source of stress factor	Descriptive Statistics	
		No. of respondents	Mean
1	Warmness of days	250	3.85
2	High price of inputs	249	3.75
3	Malaria	249	3.65
4	Population pressure	247	3.64
5	Early cessation of rain in farming season	249	3.62
6	Delay of rains in farming season (late onset)	249	3.50
7	Insufficiency of rain in farming season	250	3.49
8	Lack of agricultural inputs	248	3.36
9	Limited availability of farm land	246	3.32
10	Too early rainfall which is not sustainable	248	3.26
11	Irregular and unpredictable rains	250	3.26
12	Low level of soil fertility	249	3.14
13	Drought	247	3.13
14	Occurrence of heavy rain in few days	248	3.03
15	Flood	249	2.93
16	Lack of oxen	249	2.69
17	Lack of credit	249	2.29
18	Lack of market to purchase inputs and sell outputs	250	2.17
19	Lack of vehicle road	246	2.16
20	Lack of extension services	245	1.80

Source: Own field research data

Among individual stressors, warmness of days (which is a climate factor) topples as the leading source of concern among all of the 20 stress factors. Then follows higher price of inputs from that of non-climate factors. Apart from warmness of days, malaria and early cessation of rainfall (both climate factors) are ranked among the most top five important factors. Whereas, population pressure from the non-climate factors is positioned in the top five stressors. This result is found consistent with information obtained from focus group discussions, where farmers identified highly fluctuating behavior of rainfall, rising temperature, increasing prices of inputs (fertilizer and seed) and population growth as the most pressing problems in their livelihoods. But, farmers in the focus group discussions also identified unemployment[64] among the educated youth as one of the grave problems in recent years.

Non-climate variables such as lack of oxen, credit, market, vehicle road and extension services, which are believed to be the most important variables in farming, are pushed down rather to the bottom of the ranking. This might not be surprising given the government's commitment to the expansion of extension services, rural roads, markets and credit services. For instance, despite lack of information about the quality of services (beyond the scope of this study), which should need further research, farmers in each Kebele are provided with extension services by three agricultural development agents each skilled either in crop, livestock or natural resources development. Therefore, in general climate factors seem to be more the source of concern for farmers as compared to the non-climate stress factors.

Concluding Remark

It is argued in chapter 2 that while the discourse is hanged up at global level and is stranded in lack of consensus on the science and response actions, it is subsistence farmers in the developing world that continue to suffer the most due to their reliance on climate sensitive sectors. These farmers continue to live largely detached from the discourse. Nevertheless, they have their own bulk of knowledge that they acquired from experience and past generations. It is based on this knowledge that they first and foremost assess the local climate, and in turn govern their adaptive behavior. Understanding this bulk of knowledge is thus crucial to design practice based response mechanisms and policies that are sensitive to localities. To this end, an attempt is made in this chapter to explore local knowledge and assess its authenticity based on available scientific data.

Accordingly, local knowledge has demonstrated pronounced changes in the local climate. Through cross-examination (wherever possible) with meteorological recordings and experts' views, this knowledge system is found to be in conformity with objective data. It has offered worthwhile information on detection, attribution and impacts assessment of climate variability and change in the localities. It has provided several instances of the changes observed. It has also supported its claims through concrete examples of detection mechanisms. It is also found to attach meaning in describing the terms climate variability and climate change through the impacts that follow the specific characteristics of the changes in the local climate.

---

[64] According to farmers, graduates from secondary schools, technical and vocational training institutions, colleges and universities sit ideal at home due to inability to get employment. Either the graduates are not participating or ineffective in the farming sector as they did not develop the skills and/or lack interest in farming and rural life as they spent their life in attending schools in towns.

This knowledge system has also offered an important lesson as regard to attribution of climate variability and change, one of the major controversial topics in the discourse. In fact, it attributes the changes both to anthropogenic sources and God's will. The former cause brings it in line with what is said largely in the climate change discourse. The later cause takes it back to religious beliefs, which is regarded as culture in the materialistic view. Even though local knowledge believes in God's will and regards the changes as signs of punishment for disobeying Him, it interestingly puts weight on man's infringement of nature's stewardship given to him by God, the creator. This notion of lack of stewardship, as termed in this study, although appears to come literally under God's will, it is a cultural form of attributing the causes to human activities. Therefore, local knowledge directly or indirectly supports the idea of anthropogenic causes.

As regard to the impacts, this knowledge system demonstrated the impacts observed on the availability of water, health (new malaria appearance in the highlands), farming, and infestation of insects and weeds. In addition to these impacts, it has also revealed some psychosocial impacts that are rarely, if any, reported in the literature. The psychosocial impacts basically refer to confusion on timing of planting and tearing the social fabrics of helping each other (the form of social capital which assists the needy and disadvantaged sections of the community). Besides these psychosocial findings, the other general impacts revealed by the knowledge system are found to confirm assertions of the discourse but with unique details and specificities when it comes to impacts at local levels. Such unique insights thus make local knowledge, as a knowledge system to be exploited to facilitate and guide policy dialogue at local levels.

Chapter Five

The Economic Impact of Climate Variability and Change on Crop Production

Introduction

In the previous chapter, besides exploration of local knowledge, various impacts of climate variability and change were discussed qualitatively based on farmers' knowledge and experience. In this chapter, the discussion changes its course towards quantitative assessment of the impacts. Since crop production is the single source of farmers' livelihoods and farmers are seriously concerned with the impacts in this particular sector, this sector is selected for further analysis. First, the analysis estimates the impacts of climate variables (rainfall and temperature) and adaptation practices on the annual gross income of crop production of the year 2009 (2001/2002 Ethiopian Calendar Year). Based on this estimates, it tries to project the impacts of climate change in the mid-21st century. It finally compares and contrasts the results of the projection with the impacts of climate variability so as to suggest to which forms and timeframe policy should pay attention when dealing with local climate effects.

5.1. The Model Framework

As shown in the figure below by Theme 2a and Theme 2b, the central part of the analysis in this chapter revolves around estimating the impacts of climate variability and change on crop production, and the effects of counteractive actions (adaptation). Ricardian approach has been used to estimate such impacts. Nevertheless, it warrants that we have to propose an alternative to the Ricardian approach due to its weaknesses discussed in chapter 2 in the context of subsistence farming.

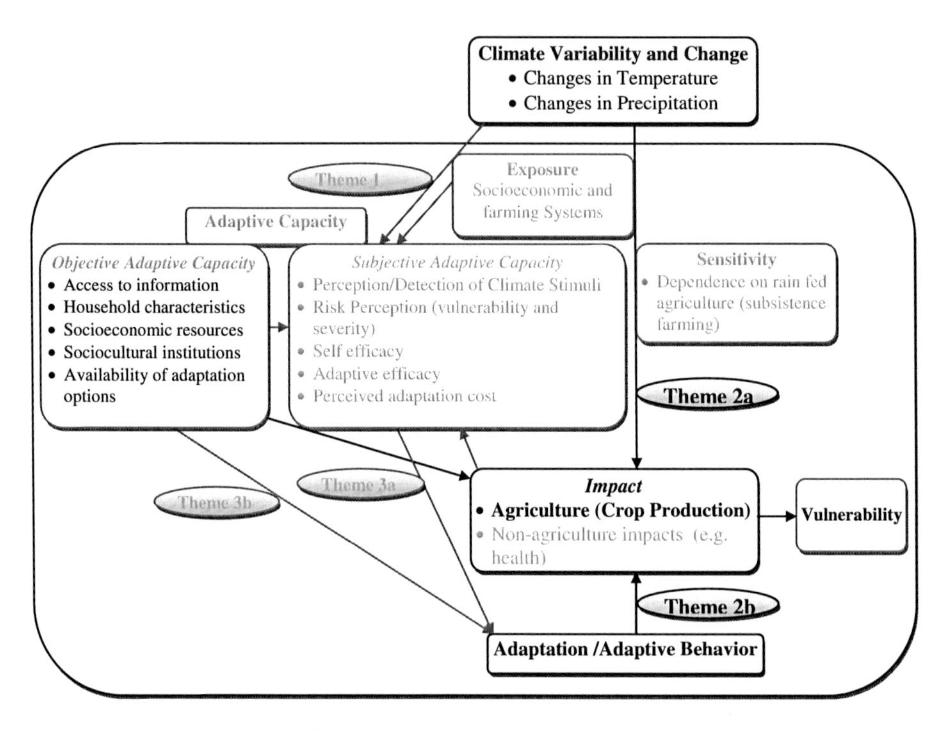

Figure 5.1. The major themes and issues addressed in this chapter (key elements are highlighted) (Theme 2a and 2b).

The alternative chosen to this particular study dwells on the usual agricultural production function described in equation 2.2 in chapter 2. Agricultural production function is one of the most popular and widely applied functions in econometrics to establish the relationship between inputs and output. In its general form, it is given as:

$$\boldsymbol{lnY = lnA + \beta_0 lnL + {}_1 lnLn + \beta_2 lnK + ln\varepsilon} \qquad (5.1)$$

Where $\boldsymbol{Y}$ is output, and $\boldsymbol{L}, \boldsymbol{Ln}$ and $\boldsymbol{K}$ are labor, land and capital inputs, respectively. $\boldsymbol{A}$ is total factor productivity (TFP) or Solow residual and $\boldsymbol{\beta_0}$, $\boldsymbol{\beta_1}$ and $\boldsymbol{\beta_2}$ are the corresponding parameters to labor, land and capital, and $\boldsymbol{\varepsilon}$ is normally and independently distributed random error that represents other factors that are not considered explicitly in the function, and $\boldsymbol{ln}$ is natural logarithm.

Such form has been widely applied to measure the contribution or marginal effect of each input on agricultural output. Climate, as an important element in agriculture, is a variable in which varying instances has been entertained in its place in the production function. Until some decades ago, econometric models neglect the specification of climate variables as exogenous factors in estimating production functions (Oury, 1965). This could be due to either or all of the following interlinked reasons.

a. Considering climate as given (given favorable climatic condition), while an enormous interest on the other manipulative economic inputs such as labor and capital.
b. Considering climate as costless input supplied by the environment and hence greater attention to cost incurring inputs.
c. Assuming climate at the random error along with other unaccounted factors.

If major climate factors such as precipitation and temperature are assumed to be nearly at their average levels or optimal levels or with the consideration that nature, social and productive systems have some level of tolerance or resilience over fluctuations on climatic conditions, standard production functions without climatic variables could be persuasive. However, in this era of concern over climate change and increasing climate variability, where natural and social resilience is threatened and farmers are suffering considerably, interest towards augmenting regression models with climatic variables has grown considerably. Since economic theory does not clearly specify the place of climate in econometric models, researchers are left to make their own decisions in the process of augmenting models with climate variables. In fact, other than this it is customary that researchers have to make several decisions on the algebraic form of a function, the choice and form of variables to be included and the technique to estimate coefficients, among other things (Griliches, 1963). When it comes to climate variables, it is not uncommon to observe a variety of variable choices and the form in which they are treated in a production function.

Based on the literature, two important empirical traditions could be constructed as to how climate variables are treated in the overall choice of variables and the form in which they are represented. With the choice of variables, one encounters mixed variables versus climate alone specification. In terms of the functional form, one also faces with linear versus quadratic terms in allowing climate variables in agricultural output analysis.

There is a wide range of empirical evidence in the literature that estimates the impacts of climate variables directly in combination with conventional inputs and other socioeconomic and environmental characteristics as given in the form of the following equation.

$$lnY = lnA + \beta_0 lnL + \beta_1 lnLn + \beta_2 lnK + \beta_3 lnZ + \beta_4 lnC + ln\varepsilon \quad (5.2)$$

Where ***Z*** stands for socioeconomic and environmental characteristics, ***C*** denotes climate variables, and others as denoted in equation 5.1.

A good example in this regard is Ricardian approach[65], where climate variables are estimated along with conventional inputs and other socioeconomic and environmental variables. Others such as Exenberger & Pondorfer (2011) regressed climate variables along with socioeconomic variables in their study of the effects of climate change in Sub-Saharan Africa. Similarly, Eckaus & Tso (1999) estimated a production function to identify the contribution of precipitation, temperature and other climatic factors along with other farm inputs on the productivity of some selected crops in China. Quiroga et al (2011) also estimated the yield response of main Mediterranean crops with respect to economic, water (irrigation), management, geographic and climatic factors. These are few examples, yet, the literature presents vast records of similar approaches.

On the other hand, others opt to estimate production function with climate variables alone by taking out the non-climate variables from equation 5.2. A classical work by Oury (1965) in the USA, and some studies such as Rowhani et al (2010) in Tanzania, and Asbjørn et al (2004) in Norway considered climate variables only (mainly precipitation and temperature).

The way in which climate variables allowed in a production function with and without economic variables seems to depend on the choice of the researcher, assumptions and objectives of the research. If more interest is on determining the effect of each of the relevant climatic variables assuming optimal utilization of socioeconomic variables, one regresses climatic variables alone on output. However, one can also note about the capability of climate variables in altering agricultural output by interfering with the physiological growth of the crop plant. To reduce the negative effects of climate variables, farmers might make various decisions on the choice and utilization of the socioeconomic inputs. Such decisions might generate the interest to analyze climate variables along with socioeconomic variables, either to control or assess the influence of socioeconomic inputs in the face of some given climatic conditions. In such assumptions, researchers appear to estimate production function by allowing climate variables along with the conventional inputs and other socioeconomic factors (as shown above in equation 5.2). As discussed in chapter 2, climate variables, however, influence agricultural output via factors of production, for example, by altering soil moisture. The same is with socioeconomic characteristics. For instance, education alters the quality of labor and hence influencing output indirectly through labor. This implies that climate variables can increase or decrease agricultural output without any additional use of

[65] In fact Ricardian model assumes quadratic terms of climate variables. For details, refer to any Ricardian approach or chapter 2.

factors of production (land, labor and capital) by influencing the productivity level of these factors. This in turn implies for factoring climate variables on total factor productivity and socioeconomic characteristics on labor.

Beyond the choice of variables, the functional form, i.e., as to how climate variables could be fitted to the equation is also another issue of interest. Due to its simplicity and theoretical support, linear approach is widely used. Quadratic[66] form is also commonly applied to capture the diminishing marginal returns. Particularly this is the norm in the Ricardian approach. Beyond other complications, using the squares of climate variables might suffer from multicollinearity. For instance, Segerson & Dixon (1998) reported severe multicollinearity in particular to the inclusion of squared climate variables that goes up to a condition index of 700,000 in their cross-sectional analysis of sensitivity of crop yield and profit to climate change in Midwestern counties of USA. Nevertheless, the issue of multicollinearity is rarely reported (whether encountered, tested or not) in many of the Ricardian studies. For example, out of 25 studies reviewed for the sake of whether they have mentioned or addressed multicollinearity, only 6 of them raised the issue. Among these studies, Kabubo-Mariara & Karanja (2006) were forced to drop several variables in their study of the economic impacts of climate change on crops in Kenya due to serious problems of multicollinearity, particularly pertaining to climate variables. Mano & Nhemachena (2006) also stated that even if they did not test for multicollinearity, they had anticipated the problem in their analysis of Zimbabwean agriculture. As a result, they had opted to use quantile regression analysis[67] rationalizing that such technique corrects the effect of outliers, and heteroscedasticity and multicollinearity as well. Still, they reported that they had dropped some variables due to multicollinearity. Similalry, García-Flecha & Viladrich-Grau (2005) found out strong correlation between temperature terms and the corresponding quadratic terms as large as 769.08 Condition Index.

It is based on these setting that a model for this study was developed. Given a kind of standard production function specified in equation 5.1, climate variables could be augmented to the production function on the basis of either or part of the following considerations.

1. Specify climate at the main function along with the conventional inputs, as discussed above.
2. Climate as the source of total factor productivity[68], so capture it at A,

---

[66] See equation 2.1 in chapter 2.

[67] Quantile regression is a model introduced by Koenker & Bassett (1978) to enhance the robustness of models, such as for outliers, in which quantiles of the conditional distribution of the dependent variable are expressed as functions of observed covariates (Koenker & Bassett, 1978; Koenker & Hallock, 2001).

[68] It is widely acknowledged that climate is one of non-conventional variables of TFP. See Rosegrant & Evenson (1995), Islam & Salim (2009), Ali et al. (2008) & Warr (2012) as few of the resources from the vast literature.

3. Climate as an error term, so constituting a portion of an unexplained or unobserved part of a function.

Theoretical considerations incline to support instances stated in number 2 and 3. It is common understanding to account non-conventional inputs such as climate in the random error term. Similarly, studies focusing on TFP relate climate not directly to unit inputs rather regard it with other terms that affect TFP. Such conceptualization warrants climate holding a dual characteristic both as a random variable and a term in TFP. As a matter of fact, rather than as a direct input, climate more importantly plays a role in decreasing or increasing total factor productivity. Assume farmers employing conventional inputs (labor, land and capital) to obtain certain level of crop yield with expectations of relatively good climate but do not know what would exactly happen to the climate at any stage of the crop growth. If climate happens to be good, farmers get good production; and conversely if it happens to be bad, production declines. But, in both cases the inputs (labor, land and capital) stay the same. Similar is the case when climate is assumed at the random term. Such is a case where climate (in the form of climate change and climate variability) lets its influence over production, being influencing total factor productivity. Particularly, given the socioeconomic characteristics of subsistence farmers, i.e., possessing small plots of land, very limited financial resources and technical level, climate variability and change largely influence total factor productivity.

Therefore, rather than the usual approach where climate is set with the conventional inputs in a function, it seems more compelling to capture climate at its role in influencing TFP and random term which leads us to capture it as a residual (unexplained part in the regression). Such framework, consequently, involves a two step regression analysis. The first step to obtain an unexplained part, i.e., the residual; and the second to analyze the effects of climate variables on the residual. Like Solow's residual, residual here refers to the part of the dependent variable that remains unaccounted by the conventional inputs.

5.2. Model Specification

5.2.1. Stage one

In the first part of the regression, output is estimated with the conventional inputs. Given the nature of subsistence farmers, it is assumed that the monetary value of crop produced in a household is determined by the level of labor, land, livestock, fertilizer and seed employed in the farming. In its logarithmic form, the equation is expressed as:

$$lnCROPROD_i = lnA + \beta_1 lnLABOR_i + \beta_2 lnLAND_i + \beta_3 lnLVSTK_i + \beta_4 lnFERT_i + {}_5 lnSEED_i + ln\varepsilon \quad (5.3)$$

Where:

- **CROPROD** is the dependent variable in monetary value (Ethiopian Birr) of annual output of eight grain crops (referred to as gross income or income onwards). The aggregate monetary value is composed based on local producer prices of each crop. Transforming the product into monetary value is required due to a mixture of crops grown by a household and the possible practice of intercropping by some farmers. The grains include teff, wheat, barley, sorghum, millet, rice, corn and haricot bean. These grains were chosen based on their enormous share in the nation's crop production. For instance, the first seven cereal crops covered about 85.94 percent of the country's total grain crop production in 2009/10 (CSA, 2010). Haricot bean is included as farmers in Rift Valley produce it as one of the major crops in the area and there is a growing tendency among farmers due to its economic benefit and resistance to intermittent rainfall.
- **LABOR** is man-labor[69] available in the household.
- **LAND** is land in hectares allotted for cultivation of the crops.
- **LVSTK** is livestock capital in Tropical Livestock Unit (TLU)[70].
- **FERT** is the amount of fertilizer used in kilograms.
- **SEED** is the amount of seed used in kilograms.
- $\beta_1, \beta_2, \beta_3, \beta_4$ and β_5 are respective parameters to be estimated.
- *ln* is the natural logarithm.
- *i* represents the household, and others (**A** and **ε**) are as stated above in equation 5.1.

5.2.2. Stage two

As discussed above, the major purpose of the first stage estimation is to obtain $lnA + ln\varepsilon$. After obtaining $lnA + ln\varepsilon$, the final regression uses this part of the income to regress it on climate variables and other socioeconomic and household characteristics. For the second stage, rearrangement of equation 5.3 yields,

$$lnA + ln\varepsilon = lnCROPROD - (\beta_1 lnLABOR_i + \beta_2 lnLAND_i + \beta_3 lnLVSTK_i + \beta_4 lnFERT_i + \beta_5 lnSEED_i) \quad (5.4)$$

[69] Labor force in the household is converted into adult male equivalent.

[70] Tropical livestock unit was used to aggregate the livestock unit of the household.

Let $\boldsymbol{lnA} + \boldsymbol{ln\varepsilon}$, be $\mathbf{r}\boldsymbol{CROPROD}$, which signifies the portion of income unaccounted by the conventional inputs specified in equation 5.3. The regression equation would thus be:

$$\begin{aligned} lnrCROPROD_i = {} & lnrA + \beta_1 lnPPT_j + \beta_2 lnMXTEMP_j + \beta_3 lnMNTEMP_j + \\ & \beta_4 DVPPT_j + \beta_5 SEX_i + \beta_6 AGE + \beta_7 EDUC_i + \beta_8 AGREXP_i + \\ & \beta_9 EXTN_i + \beta_{10} ACCREDIT_i + \beta_{11} NOSO_i + \beta_{12} DIST_i + \beta_{13} ADAPT + \\ & ln\varepsilon \qquad (5.5) \end{aligned}$$

Where:

PPT	=	Total amount of annual precipitation in millimeter.
MAXT	=	Average annual maximum temperature in degree centigrade.
MINT	=	Average annual minimum temperature in degree centigrade.
DVPPT	=	Deviation of annual precipitation in millimeter from the long term mean (20 years) and it represents one form of climate variability.
SEX	=	Sex of the head of the household.
AGE	=	Age of the head of the household in years.
EDUC	=	Education level of the head of the household in years.
AGREXP	=	Farming experience of the head of the household in years.
EXTN	=	Frequency of the household's contact with extension agents in a year.
ACCREDIT	=	Household's access to credit; a dummy which takes the value of 1 if has access and zero otherwise.
NOSO	=	Number of social organizations the household is a member.
DIST	=	Distance to market in hours.
ADAPT	=	Represents whether a household has undertaken adaptation measures, and is a dummy, which takes the value of 1 if performed and zero otherwise.

As regard to climate variability, there are a number of forms that could constitute it, such as timing of rainfall (onset and cessation), interruptions and extreme events (flood, drought etc). However, unavailability of data either in meteorological recordings or the natural absence of the events from occurrence limits the consideration of these aspects of climate in the regression. To bridge data limitedness in cases of occurrence, an attempt was made to generate data from farmers' assessment of the farming year. This is dealt in depth in section 5.4.

5.3. Estimation Technique and Econometric Issues

OLS was used to estimate the coefficients of the variables. Apparently, OLS estimators are assumed to be unbiased and efficient, as commonly tagged as Best Linear Unbiased Estimator (BLUE). The BLUE property requires the OLS to live up to certain assumptions. Reliability of the coefficients depends on the fulfillment of these assumptions. It assumes linearity of the estimators, homoskedasticity of the disturbance term, and absence of multicollinearity among the independent variables. Both the models are then subject to the diagnosis of these classical assumptions of OLS in the following ways.

a. Linearity in the functional form

It is assumed that the dependent variable is a linear function of the explanatory variables. This assumption directly implies the specification and the functional form of the model. This assumption could be violated if one or more of the following are encountered (Gujarati, 2004).

- Omission of relevant variable (s) (under fitting the model)
- Inclusion of an unnecessary variable(s) (over fitting the model)
- Specification of incorrect functional form
- Measurement errors

To insure that the models retain the linearity assumption and are free from specification error, Ramsey's RESET[71] was carried out. RESET is a popular test used to identify whether the explained variable is linear on the explanatory variables and if there exists misspecification in a model. It has the null hypothesis that the model is free of omitted variables against the alternative hypothesis of model misspecification. A non-significant test result is thus an indication that the model is fit for linearity.

b. The Assumption of Homoskedasticity

One of the basic assumptions in OLS is homoskedastic distribution of the error term. The error term is presumed to have zero mean value and constant variance around its mean. If the error term fails to have constant variance, heteroskedasticity occurs. In fact, the presence of heteroskedasticity does not cause bias or inconsistency in the estimators. But OLS is no longer BLUE, i.e., the standard errors are invalid or no longer valid for constructing t statistics and confidence intervals (Wooldridge, 2002b; Salvatore & Reagle, 2002). Therefore, heteroskedasticity tests are necessary in a regression analysis, particularly in cross-section data, where it usually suffers from the problem (Verbeek, 2004). Heteroskedasticity could be inspected visually[72] or through certain statistical tests, such as Breusch-Pagan/Cook-Weisberg test and White's test. Breusch-Pagan/Cook-Weisberg test was carried out in this study.

---

[71] RESET stands for regression specification error test. For detailed discussion see, for example, Ramsey (1969); Wooldridge (2002a), Chapter 6; Gujarati (2004), Chapter 13; http://www.ats.ucla.edu/stat/stata/webbooks/reg/chapter2/statareg2.htm, accessed 28/07/2012.

[72] By plotting the residuals against fitted (predicted) values.

c. The Assumption of Normality

The other assumption with regard to the error term is normality. The property of BLUE requires a normal distribution of the error term. In fact, violating the assumption of normality might not be as such a serious problem given the assumption of asymptotic normality in sufficiently large sample size and with the condition that other assumptions hold true (Wooldridge, 2002b; Verbeek, 2004). Despite this, meeting normality criteria in any case increases the consistency of the estimates. Visual and statistical tests could be employed to test normality. Since visual tests tend to be subjective, Jarque-Bera normality test was taken for the test. Jarque-Bera test presupposes the null hypothesis of normally distributed error terms against the non-normal distribution of the alternative hypothesis, which in this case smaller and non-significant values granting evidence for normality.

d. Multicollinearity

Another important diagnostic test in regression analysis is the issue of multicollinearity. Classical linear regression models assume no multicollinearity among the explanatory variables. In real econometric estimation, multicollinearity is a relative term. Any multiple regression suffers from multicollinearity to some level, unless the independent variables are uncorrelated at all, which is hardly realistic (Dougherty, 2001). Gujarati (2004:354) states that "as long as multicollinearity is not perfect, estimation of the regression coefficients is possible...". Therefore, multicollinearity is the matter of degree, not kind; and it happens to be a problem when the degree becomes high or nearly perfect (Kmenta, 1986 in Gujarati, 2004; Dougherty, 2001). Actually, the presence of multicollinearity does not affect the BLUE property of OLS, but it increases the variance and standard errors of the estimates. It also makes very difficult to discern the separate effect of each of the intercorrelated variables on the dependent variable (Maddala, 1992; Wooldridge, 2002b; Salvatore & Reagle, 2002; Gujarati, 2004). To rule out multicollinearity or take a remedy in the case of its presence, diagnostic tests were carried out. Mostly used post estimation tests, namely, variance inflation factor (VIF) and/or tolerance were/was used to measure the level of multicollinearity. VIF indicates how much inflation in the standard error is caused by collinearity, and tolerance, on the other hand, shows the amount of collinearity that a regression can tolerate (http://www.ats.ucla.edu/stat/stata/webbooks/logistic/chapter3/statalog3.htm, accessed 28/07/2012). As a rule of thumb, VIF greater than 10 or tolerance index less than 1 are taken as indications of the problem (Adkins, 2009; Gujarati, 2004, http://www.ats.ucla.edu/stat/stata/webbooks/logistic/chapter3/statalog3.htm, 28/07/2012). This rule of thumb is followed in this study as well.

5.4. Model Estimation and Discussion

As noted earlier, the main purpose of the first regression is to obtain the residual part of the households' income from cereal crops. Hence, it might not be necessary to go in detail on discussing the results beyond a brief explanation. However, as the second regression is reliant on the output of this regression; more emphasis thus lays on diagnostic tests rather than the output discussion. Testing the assumptions of OLS helps to get reliable outputs for the second stage.

Table 5.1 presents the regression results along with the diagnostic tests. The model passes all of the OLS assumptions with the exception of the assumption of homoskedasticity. Jarque-Bera normality test shows that the error term is normally distributed with a value of 0.857 ($\chi^2 = 0.308$). Similarly, RESET, the test for linearity, reveals a non-significant value of 2.06, which satisfies the linearity assumption of the model. Collinearity diagnostics[73] also shows no sign of multicollinearity with VIF values much less than the critical cut point of 10.

As regards to the assumption of homoskedasticity, Breusch-Pagan/Cook-Weisberg reveals a significant chi-square value of 7.06[74], which fails to reject the absence of heteroskedasticity. Therefore, the regression was estimated with White's correction procedure. In such cases, it needs to be noted that the reported standard errors are that of robust standard errors (heteroskedasticity-consistent standard errors or White standard errors) than the usual (normal) standard errors customarily reported in homoskedastic regression models. Along with this, since STATA does not report adjusted R-squared in the application of White's correction method, only R-squared is reported in the table below and elsewhere in other tables.

[73] See the details of the analysis in Appendix 5.

[74] See the details of the analysis in Appendix 5.

Table 5.1. OLS estimation of the impact of conventional inputs on annual income from cereal crops

Dependent Variable: CROPROD (Households annual income from cereal crops' production)

<table>
<tr><th>Variable</th><th>Coef.</th><th>Robust Std. Err.</th><th>t</th><th>P>|t|</th></tr>
<tr><td>LABOR</td><td>0.11995</td><td>0.062307</td><td>1.93</td><td>0.055*</td></tr>
<tr><td>LAND</td><td>0.582653</td><td>0.071408</td><td>8.16</td><td>0.000***</td></tr>
<tr><td>LVSTK</td><td>0.142521</td><td>0.035945</td><td>3.96</td><td>0.000***</td></tr>
<tr><td>FERT</td><td>0.175706</td><td>0.055986</td><td>3.14</td><td>0.002***</td></tr>
<tr><td>SEED</td><td>0.155266</td><td>0.055064</td><td>2.82</td><td>0.005***</td></tr>
<tr><td>Constant</td><td>6.942603</td><td>0.261911</td><td>26.51</td><td>0.000***</td></tr>
<tr><td>R-squared</td><td colspan="4">0.5807</td></tr>
<tr><td>F - statistics</td><td colspan="2">70.33</td><td colspan="2">Prob > F = 0.0000</td></tr>
<tr><td>Diagnostic test</td><td colspan="4"></td></tr>
<tr><td>Jarque-Bera normality test</td><td>0.857</td><td colspan="3">χ^2 = 0.3085</td></tr>
<tr><td>RESET test</td><td>2.06</td><td colspan="3">Prob > F = 0.1062</td></tr>
<tr><td>Multicollinearity</td><td colspan="4">Mean VIF 1.30, the detail is shown in Appendix 5</td></tr>
<tr><td colspan="5">NB: *** significant at 1% level, * significant at 10% level</td></tr>
</table>

Source: Own field research data

With regard to the results, few words could be said before passing to the main part of the regression. As shown in Table 5.1, all of the five variables significantly affect the households' total income from cereal crops. These variables all together explain about 58.07 percent of the variation on the dependent variable. The coefficients of the variables are all positive and statistically significant. The positive sign of the coefficients is an indication that the estimates are in line with predictions of economic theory. In fact, four of the variables with the exception of labor are significant at 1 percent level. Labor turns significant when the probability level rises to 10 percent, and this may be an indication for surplus inclination of labor.

Of the variables, land takes the higher influence, i.e., 1 percent increase in land hectare leads to an increase of annual income by 0.58 percent. Such result is an evidence of the level of dependence of subsistence farmers on mining natural resources. Since technological application is low or negligible among such farmers, crop production is dependent to a great extent on land resources to meet food requirements of the burgeoning population. In connection to this, Aberra (2011) indicated that land area expansion in Ethiopia was the major cause for production gain in between 1975 and 1992, which by implication would have serious consequences on local climate and environmental sustainability. Such enormous dependence on land undermines local environment and its sustainability as far as new and marginal lands continue to come under cultivation. This is typically consistent with farmers' views discussed in the previous chapter where they link the changes in the local climate to deforestation induced by reclamation of new land outlets for agricultural purposes. Given the scarcity of land resources in the country and the environmental implications of expansion of farming land, the results appeal to a big call for the need to increase productivity from yield improvement or input intensification.

Having said this, it is time to turn to the second stage of the regression, where unaccounted part in the variation of the households' total income by conventional inputs is explained by climate factors and socioeconomic variables that influence the efficacy of crop production. Alike the previous regression, the first step pays attention to the diagnostic tests of the model as to whether it satisfies OLS assumptions, and consider remedial actions in case of breaching the assumptions. Accordingly, the model is found to satisfy the assumptions of linearity and normality (see the statistical figures in Table 5.2). Multicollinearity diagnostics as measured by VIF also shows no cases of intolerance among the variables (see statistical figures in Appendix 6). However, with chi-square value of 9.87 and probability value of 0.0017, it was not possible to rule out the presence of heteroskedasticity. Therefore, the regression was run with heteroskedasticity-consistent standard error estimator.

The results of the parameter estimates are presented in Table 5.2. All of the climate variables, i.e., precipitation, maximum temperature, minimum temperature and deviation of precipitation, and three of the socioeconomic variables, namely, age, number of contacts with extension agents and adaptation are found to be important factors in their effect on income. The overall fit of the model is significant with R-squared of 52.58 percent. Among the climate variables, precipitation and minimum temperature have significant and positive impacts. On the contrary, deviation of precipitation and maximum temperature influence income significantly and negatively.

Income increases by 0.88 percent from an increment of annual precipitation by one percent and a 1 percent rise in minimum temperature boosts it by 2.33 percent. On the other hand, a 1 percent increase in maximum temperature reduces income by 3.2 percent. Similarly, divergence of precipitation from its long term mean value by 1 millimeter reduces income by 0.08 percent. This shows that deviation of precipitation from its long term mean either way, i.e., upwards or downwards, is harmful for cereals' production. Such result is consistent with Aberra's (2011) study in Ethiopia, where he reported diminishing response of crop production to rainfall divergence from its long term mean. In overall, these results of climate variables might not be as such surprising given the high dependence of subsistence farming on climate. It has been well documented that precipitation is the most important single factor that determines agricultural production in Ethiopia. The figure presented in chapter 1 (Figure 1.1) could suffice this assertion. However, in addition to precipitation, temperature has shown its paramount importance in determining crop production. The importance of temperature could be more tantalizing in the face of increasing trends of both maximum and minimum temperature. From this point of view, temperature plays a mixed role: a boosting and a detrimental effect. Increasing maximum temperature has negative effects on income, which contrasts benefits from increasing minimum temperature. Particularly, the effect of temperature would be more noticeable in the long run under climate change at some point in the future (see the discussion in section 5.5).

Table 5.2. OLS estimation of the impact of climate variables

Dependent Variable: rCROPROD (Households annual income from cereal production, unaccounted portion in the variation of income)

Variable	Coef.	Robust Std. Err.	t	P>|t|
PPT	0.885846	0.121261	7.31	0.000***
MXTEMP	-3.20789	1.315193	-2.44	0.015**
MNTEMP	2.332935	0.379852	6.14	0.000***
DVPPT	-0.00085	0.000468	-1.82	0.070*
SEX	0.0269	0.067814	0.40	0.692
AGE	-0.00689	0.004083	-1.69	0.093*
EDUC	-0.00309	0.009255	-0.33	0.739
AGREXP	0.000999	0.003923	0.25	0.799
EXTN	0.003143	0.001668	1.88	0.061*
ACCREDIT	0.008687	0.053957	0.16	0.872
NOSO	-0.02877	0.034613	-0.83	0.407
DIST	-0.00527	0.046664	-0.11	0.910
ADAPT	0.086885	0.046759	1.86	0.064*
Constant	5.994904	3.944072	1.52	0.130
R-squared	0.5258			
F - statistics	20.28	Prob > F = 0.0000		
Diagnostic test				
Jarque-Bera normality test	4.155	χ^2 = 0.1252		
RESET test	2.03	Prob > F = 0.1106		
Multicollinearity	Mean VIF 2.77, the detail is shown in Appendix 6			
NB: *** significant at 1% level, ** significant at 5% level, * significant at 10% level				

Source: Own field research data

When we turn to non-climate variables, one year increase in age decreases income by 0.68 percent. Such negative results have been observed in a number of studies, particularly in those analyzing the technical efficiencies of farmers. In fact, there are also mixed results in the literature pertaining to such factors as education, age and experience. However, from theoretical perspectives, old age is more likely linked with production inefficiency. The arguments behind this rests on the fact that old farmers are usually resistant to the application of new methods and they might be less capable both physically and technically to carry out the cumbersome activities of subsistence farming, which eventually affects production negatively. For instance, Tiamiyu et al (2010) found a negative relationship between age and productive efficiency indices.

The coefficient for households' contact with extension agents is positive and statistically significant. Increasing the annual number of extension contact by one increases income by 0.31 percent. Extension services[75] have been found beneficial for agricultural production in other similar studies as well (Mano & Nhemachena, 2006; Molua & Lambi, 2006). Since extension services facilitate information flow on productive technologies and markets, the positive and significant results could be attributed to such reasons. Particularly in the context of climate variability and change, extension services help to provide information on weather and adaptation options to prepare farmers to adapt to adverse climatic conditions.

Another interesting result is the benefit of carrying out adaptation strategies to climate variability and change. Performing adaptive strategy is found to increase income by 8.68 percent than failing to do so. In line with this, Molua & Lambi (2006) found out that implementation of adaptation strategies boosts production among farmers in Cameroon. It implies that adaptation has the potential to offset the adverse effects of existing climatic conditions.

The other non-climate variables, namely, sex, experience and credit are found to contribute positively but are statistically insignificant. On the other hand, education, participation in social organizations and distance to market are negative even though they prove to be statistically insignificant.

5.5. Projecting the Impacts of Climate Change and Weighing the Impact of Climate Variability

Based on the parameter estimates obtained from the previous regression, let us now turn to simulation of the impacts of climate change. The usual trend to measure the impacts of climate change is to take the values of precipitation and temperature at some point in the future while keeping climate variability constant, i.e., the same as the base year (Reilly et al., 2001) along with controlled socioeconomic variables. So is the

[75] Sometimes comparisons are difficult as measurement of the variable is different in many studies. This is a case also in other variables such as age, education etc.

approach here as well. With this in mind, the impacts are analyzed for the mid-21st century that runs from 2046-2069 under greenhouse emission scenario of A1B. Projected values of precipitation and temperature from three Global Climate Models (GCMs), namely MRI-CGCM2.3.2, GFDL-CM2.0 and CGCM3.1, were considered. Such data was obtained from the World Bank's Climate Change Knowledge Portal, which provides downscaled data for Ethiopia and other countries.

Due to lack of downscaled data pertaining to the highlands of Ethiopia, projected climate for the overall country was considered for forecasting the mid-century impacts. In fact, downscaled data to the highlands or specific areas could have increased the accuracy of the results. Nevertheless, the overall projected data of Ethiopia still sheds some light or could provide a rough estimation. The results might be under or overestimated, as temperature in the highlands is smaller than the overall country where lowlands with warmer temperature are also included. On the contrary, the results might be underestimated due to the fact that precipitation in the highlands is relatively larger than the corresponding lowlands. But, an assumption could be still taken that the percentage changes of temperature and precipitation in the highlands would be relatively comparable with the percentage change nationally.

Simulation results show that warmer temperature has mixed effects. Increasing maximum temperature has negative effects on income, which contrasts a benefit from increasing minimum temperature. Whereas there will be a benefit if precipitation increases. Simple arithmetic computation puts the mid century effects of climate change to have positive effects on income (a change of +20.87 percent under MRI-CGCM2.3.2 model, +17.91 percent under GFDL-CM2.0 model and +14.44 percent under CGCM3.1 model). These positive results are consistent with findings by Maddison (1998) and Deressa et al (2007). Maddison (1998) very interestingly found out that only Ethiopia incurs welfare gains from a centigrade increase of temperature among 26 poorest countries (including African countries). Similarly, Deressa (2007) found out a positive net revenue returns from projected climate (temperature and precipitation) for the year 2050. Such positive effects would be attributed to the country's relatively cooler climate which it inherits from its highlands. Ethiopian highlands exceptionally enjoy a temperate climate as compared to the rest of African countries. It is in this country that almost half of African highlands are located (Gryseels & Anderson, 1983).

Table 5.3. Projected income change in the mid-century

Model	Percentage change in income
MRI-CGCM2.3.2	+20.87
GFDL-CM2.0	+17.91
CGCM3.1	+14.44
Average	+17.74

Source: Own projection

From farmers' behavioral perspectives, there are also other reasons that envisage the effects of climate change in the mid-21st century to be positive. In the focus group discussions, farmers stated that they have been resorting into crop varieties and species that have been cultivated in the lowlands, which now turn to be suitable to the highlands due to increased temperature. Increasing corn and goaya[76] cultivation in Yilmana Densa Wereda, rice in Fogera Wereda and haricot bean in some areas of the central Rift Valley could be good examples.

Farmers in Yilmana Densa Wereda revealed that they are inclined more to the production of corn varieties which seem to suit to warming climate. They have also introduced goaya, as a new crop to their farms, to cope up with rainfall problems. Agricultural change in Fogera Wereda is another example how farmers adopt crops that suit to changing climatic conditions. Before the introduction of rice in Fogera Wereda in the 1980s, there was a problem of food security where the plains stay drenched under water during the rainy season. Coupled with suitable temperature conditions for growing rice, those water lodging plains have been transforming into rice farms in the past few decades. The Wereda now is tagged as a surplus producing area. Similarly, faced with intermittent rainfall, farmers in Central Rift Valley are also gradually inclining towards the production of haricot bean. The production of this crop has been further stimulated by its demand in domestic and international markets. Such visible agricultural changes that go parallelly with climate change could offset the impacts and might also enhance production. As far as highland farmers are able to adopt lowland crops, climate change, as understood by shifts in the mean values of climate, might not have a havocking shadow to undermine crop production in the highlands in middle of this century.

But, is this all about climate change which shows positive outcomes for crop production in the highlands of Ethiopia? Does it mean that there is less to worry by mid-21st century in these areas? To answer these questions, let us see the other aspect of climate - climate variability. As indicated earlier, studies assume constant climate variability in

[76] Goaya is a local name for the crop lathyrus sativus, also called as Indian pea or chickling pea or grass pea. The crop is well known to its drought tolerant behavior.

forecasting the impacts of climate change (Reilly et al., 2001). Because of this, the effect of climate variability quite usually remains invisible. However, for countries like Ethiopia which receive relatively adequate amount of rainfall, climate variability is more important than climate change per se. This is typically a concern among farmers as elaborated in the previous chapter.

In the above regression, it is shown that deviation of rainfall from long term mean, as one of the measures of climate variability, has negative effects. Now let us turn to another aspect of climate variability, which farmers regard it critical. Beyond deviations from long term mean, climate variability could take various forms such as fluctuations in the timing of rainfall and occurrence of extreme events. As there is lack of meteorological data in such aspects, information gathered from farmers (through focus group discussions) was used to establish a measurement of climate variability.

In focus group discussions, farmers were asked about the onset and cessation of rainfall (if there was pronounced delay/early appearance, early/late cessation or was on time) and the overall performance of rainfall in the crop calendar. Their answers as regard to onset is similar among all of the study sites which they stated that there was no as such pronounced delay or early commencement of rainfall. However, framers living in two of the study sites expressed a marked early cessation of rainfall. This information was transformed into a dummy variable (0= there was no problem on cessation; and 1 = there was a problem on cessation /early rainfall cessation/) and was inserted into the regression equation to see whether this has effect on income so as to illustrate another aspect of climate variability's impact. In addition, farmers had rated the overall performance of rainfall in the year. However, this rating was not used in the regression due to two reasons. The first is multicollinearity, where inclusion of this new variable enforces the regression to suffer from collinearity. Secondly, the overall performance of rainfall could be captured redundantly by the sum of precipitation which is already included in the regression.

While running the regression with the inclusion of rainfall cessation (symbolized as CESS in the model) along with all the variables specified in equation 5.6, intolerable amount of collinearity was encountered. Rainfall deviation was identified as the source of the problem and was dropped from the regression. The model then passed the diagnostic tests with the exception of heteroskedasticity, in which case the remedial robust standard error is reported.

Table 5.4. OLS estimation of the impact of climate variability

Dependent Variable: rCROPROD (Households annual income from cereal production, unaccounted portion in the variation of income)

Variable	Coef.	Robust Std. Err.	t	P>|t|
PPT	0.554865	0.139128	3.99	0.000***
MXTEMP	-2.89879	1.304679	-2.22	0.027**
MNTEMP	2.524151	0.424281	5.95	0.000***
CESS	-0.21428	0.117828	-1.82	0.070*
SEX	0.0269	0.067814	0.40	0.692
AGE	-0.00689	0.004083	-1.69	0.093*
EDUC	-0.00309	0.009255	-0.33	0.739
AGREXP	0.000999	0.003923	0.25	0.799
EXTN	0.003143	0.001668	1.88	0.061*
ACCREDIT	0.008687	0.053957	0.16	0.872
NOSO	-0.02877	0.034613	-0.83	0.407
DIST	-0.00527	0.046664	-0.11	0.910
ADAPT	0.086885	0.046759	1.86	0.064*
Constant	7.012036	3.919717	1.79	0.075*
R-squared	0.5258			
F - statistics	20.28		Prob > F = 0.0000	
Diagnostic test				
Jarque-Bera normality test	4.155		χ^2 = .1252	
RESET test	2.03		Prob > F = 0.1106	
Multicollinearity	Mean VIF 3.21, the detail is shown in Appendix 7			
NB: *** significant at 1% level, ** significant at 5% level, * significant at 10% level				

Source: Own field research data

As shown in Table 5.4, early cessation of rainfall significantly and negatively affects income. Early withdrawal of rainfall reduces 21.4 percent of the income as it directly affects the flowering and maturity stage of crops. This clearly shows that even one shot of a form of climate variability could wither a significant portion of crop production, and hence income. This result is typically in line with farmers' assessment of climate variables. As it was discussed well in the previous chapter, farmers are more concerned with the timing and irregularity of rainfall distribution (onset, cessation and intermittent behavior of rainfall) than the amount itself. These characteristics of rainfall obviously symbolize climate variability. Even with the awareness of the decreasing trend of rainfall, farmers are more concerned with these characteristics of the climate and attach dire consequences to these characteristics when it comes to crop production. Very interestingly, farmers also rated early cessation as one of the top five sources of stress from a list of 20 climate and non-climate factors (see Table 4.6 in chapter 4). According to the rating, early cessation of rainfall takes the first position from the various characteristics of rainfall labeled as sources of concern. The result of the regression analysis thus supports the views of farmers where climate variability is detrimental than climate change, which in this study proves to affect crop production rather positively in the highlands in the mid-21st century. Since climate variability is assumed to be constant while measuring the impact of climate change, it follows that climate variability (as far as it happens in the features stated above or other forms) will also be dismal in the mid-21st century. Therefore, climatic conditions in the form of climate variability seem to be damaging crop production among farmers in the highland areas of Ethiopia, and farmers are literally more vulnerable to climate variability than climate change.

However, if climate change lives up to the thesis that it alters the underlying elements that define climate variability and thus induces changes in the behavior of climate variability (Sperling and Szekely, 2005; Kandji et al., 2006; Washington et al., 2006), this could be regarded as sufficient condition to envisage climate change casting a gloomy prospect on crop production through climate variability but not in its mean climate values as conceived customarily. Identifying this complex nature of climate change, would thus be a huge task for the scientific world.

Concluding Remark

In this chapter, an attempt is made to project the impacts of climate change in the mid-21st century in the highlands of Ethiopia based on the monetary value of crop produced in 2009 (2001/2002 Ethiopian Calendar Year). Accordingly, it is found that farmers in the highlands of the country will see positive effects in the middle of this century. Nevertheless, climate variability as represented by rainfall's deviation from its long term mean and early cessation shows to carry adversary effects. Such negative effects of climate variability may wither up a considerable portion or much of farmers' crop production depending on the strength and the scope of its manifestations. Consequently, endangering food security and weakening farmers' capacity to purchase inputs such as fertilizer and improved seed out of the sale of some portion of the product. In fact, adaptation might play some role in lessening the negative impacts. However, the results show that the impacts are disproportionate to the stabilizing capacity of already implemented adaptation strategies by farmers. Therefore, climate variability as reflected by some of its forms in the current climate will be more damaging than climate change, which rather reflects positive effects in the highlands of Ethiopia.

Chapter Six

Adaptive Behavior among Farmers

Introduction

Regression analysis in the previous chapter has shown changes in the individual variables of climate having mixed effects on crop production. But, total effect of climate change is projected to bring about positive outcomes. On the other hand, climate variability is found to carry adversary impacts. What makes climate variability more important is not merely its significant impact but also its immediacy as a problem that farmers face currently.

To reduce the negative consequences of climate variability, farmers are supposed to undertake various adaptation strategies. The regression analysis in the previous chapter which revealed the positive effect of adaptation to crop production also strengthens this expectation. Despite this, some farmers failed to undertake adaptive strategies while others did. Exploring what lies beneath such behavior would thus be the purpose of this chapter. Hence, this chapter is exploratory in nature which attempts to identify key issues and variables that help us better explain adaptive behavior.

6.1. Model Framework

As elaborated in chapter 2, adaptation research is mainly guided by two sets of theories - adoption of agricultural innovations and Protection Motivation Theory (PMT). Conceptual frameworks from these theories are used to address the issue of adaptation in this chapter. While PMT helps to cover the psychological (cognitive) aspect of adaptation, agricultural innovations model covers the socioeconomic and institutional aspects of adaptation.

The knowledge of climate change and variability and subsequent behavioral responses at individual level are more of the products of perceptual process. This is particularly true among farmers in developing countries, where basic climate information and forecast services are very limited. From this perspective, perceptual factors play a primary role in understanding the climate and guiding behavioral response in individuals. On the other hand, socioeconomic factors intervene as enabling factors mainly to translate perceptually conceived adaptive behavior into actual behavior. This study thus deals the problem from these two compelling lines of approaches - PMT (psychological perspective) and adoption of agricultural innovations (SEI perspective).

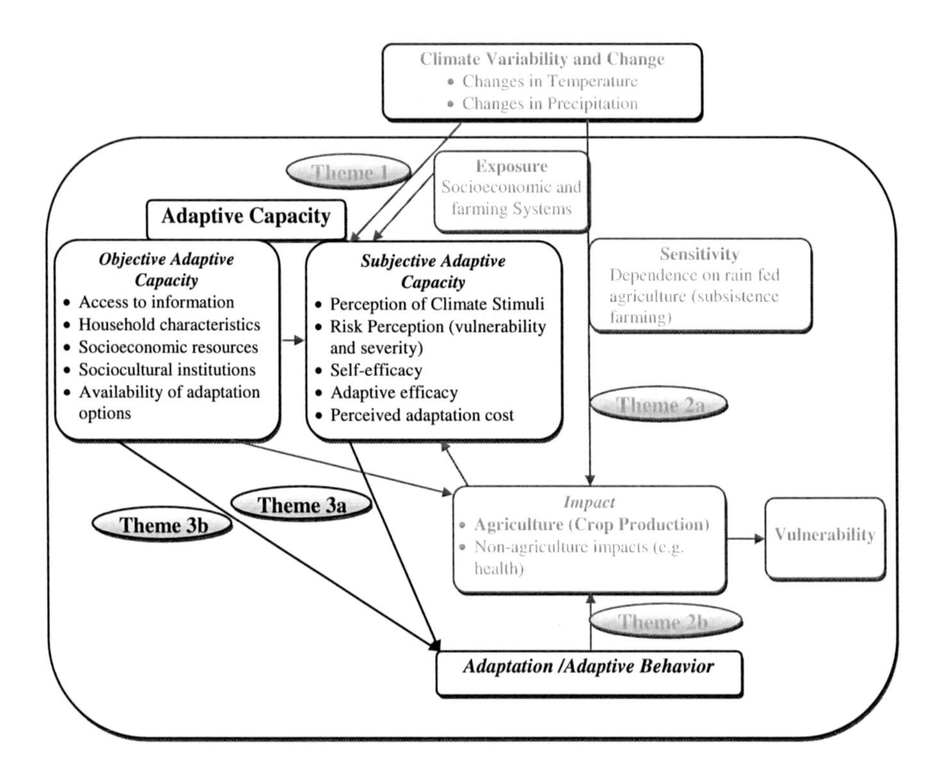

Figure 6.1. The major themes and issues addressed in this chapter (key elements are highlighted) (Theme 3a and 3b)

Pertaining to the central tenet of the approaches, the explanatory variables in PMT and adoption of agricultural innovations are different. As shown in Theme 3a above in Figure 6.1, PMT encompasses cognitive factors. On the other hand, the latter (adoption of agricultural innovations) utilizes socioeconomic factors depicted under Theme 3b in the same figure. Despite this, the dependent variable, namely adaptation or adaptive behavior, stays the same in both of the approaches.

6.1.1. Framing Adaptive Behavior from Psychological Perspective

As discussed in chapter 2, PMT has been adequately applied in analyzing individual behavioral response to environmental concerns and health risks (Menzel & Scarpa, 2005). Despite this, the application of PMT or the role of psychosocial factors in the context of climate change and variability has been given little attention (Grothmann & Patt, 2005). Because of this, measurement methods of the variables and the application of PMT is yet an emerging area of research when it comes to adaptation studies. In this study, the main variables of PMT are maintained and operationalized by fitting into the study's context.

Similar to Grothmann & Patt's[77] application of PMT (2005), households' behavioral adjustment (adaptive behavior) to climate variability and change is framed as the function of five cognitive or perception factors: perception of vulnerability (susceptibility), perception of severity, self-efficacy, perceived cost and adaptive efficacy.

$$ADAPT = f\,(PV, PS, AE, SE, PC) \qquad (6.1)$$

Where:

ADAPT is the status of adaptation; PV, perceived vulnerability; PS, perceived severity; AE, adaptive efficacy; SE, self-efficacy and PC, perceived cost.

Perceived vulnerability measures the perceived level of the likelihood of crop farming's susceptibility to harm or experiencing possible damages from mostly observed climate problems in the localities (as proxy to climate variability and change) in the absence of adaptive behavior. Climate problem here characterizes actual climatic predicaments mostly experienced by farmers. The respondents were not given hypothetical examples of climate change and variability with the concern that hypothetical instances could mask real situations on the ground and might lead to unauthentic behavioral responses. Either extreme cases of climate were not included in the statements as it was anticipated how these problems could be rated. For these reasons, measurement of perception factors were made to root on actual happenings of the climate with additional reference to household's response to questions related to the trend of rainfall and temperature in the prior part of the questionnaire (for details see Appendix 11). **Perceived severity** measures the perceived level of harm or damage that crop farming may face if or being exposed to climatic problems in the localities in the absence of adaptive behavior.

[77] Refer chapter 2 for details and further arguments.

Self-efficacy measures the perceived ability in terms of technical skills or expertise of the households to carry out adaptive strategies. In fact, self-efficacy may not only refer to skills but also perceived costs or resources required to accomplish the strategies. Originally in PMT's theoretical framework, self-efficacy refers to the skills of an individual to perform an adaptive behavior. Later with the adoption of the model in other non-health fields, it has been extended to include resources (e.g. financial capacity) required to perform the behavior. This is typically true in adaptation to climate variability and change where farmers very importantly assess their resource capacities to implement adaptation strategies. Thus, alike Grothmann & Patt (2005), self-efficacy is constructed to measure only skills, and resources are measured by their own through perceived cost. **Perceived cost** thus measures the household's level of confidence to afford the costs of resource endowments (financial resource) that are required to carry out adaptive strategies. **Adaptive efficacy** measures perceived effectiveness of adaptive actions in helping to reduce risks on crop farming from climate problems that are quite usually experienced by farmers.

It is assumed that when perceived vulnerability and perceived severity are high, households are expected to develop the sense of experiencing a considerable level of risk of losing the annual crop production. Within such risk perception, if households have developed the sense of confidence in self-efficacy and adaptive efficacy, the propensity of adaptation increases given that perceived costs of implementation are low or affordable.

The equation above (equation 6.1) and measurement of the variables here shares the feature of any other models which simplify the complex reality. Reality in the world apparently comprises a number of complex and interacted factors that cannot be entirely captured by empirical models. Particularly, this is a case in psychological process where abstract factors are involved. Similarly, the variables in this framework measure the complexity of cognitive process assuming simplification of reality.

In empirical application of PMT, the variables have been measured by behavioral statements that use Likert scales. Likewise, Likert scale was used to measure the variables in this study. Household heads were given statements[78] that are believed to stand for each of the variables. The statements were composed of five point Likert scale ranging from "not vulnerable at all (1)" to "extremely vulnerable (5)" in the case of perceived vulnerability; "not harmful at all (1)" to "extremely harmful (5)" in the case of perceived severity; "ineffective (1)" to "effective (5)" in the case of adaptive efficacy; "unconfident (1)" to "confident (5)" in the case of self-efficacy and perceived cost.

[78] Great care was taken and repeated tests were carried out before the final version of these statements was finalized. Because of time pressure, the complexity of situations due to the involvement of several adaptation strategies and conceptual similarity among the measurement statements, one statement for each variable was constructed.

6.1.2. Framing Adaptive Behavior from SEI Perspective

Analyzing adaptive behavior based on socioeconomic and institutional variables is straightforward. It follows similar approach to adoption decisions of agricultural technologies at micro-levels. As explained in chapter 2, both theoretical and empirical works also layout adaptation at local levels being determined by various SEI factors such as finance, technology, institutional and infrastructural services. In precise terms, Smit & Wandel (2006:287) state that:

> *At the local level the ability to undertake adaptations can be influenced by such factors as managerial ability, access to financial, technological and information resources, infrastructure, the institutional environment within which adaptations occur, political influence, kinship networks, etc.*

Following this, adaptive behavior from SEI perspective is analyzed as a function of an array of various socioeconomic and institutional variables pertaining to the households. With this premise, several variables might be generated. However, the variables included in this study were selected based on theory and empirical works in related fields. In addition, care was taken to include those variables deemed important in the propensity of adaptation in the context of subsistence farming. Adaptation is thus analyzed as a function of the following variables:

ADAPT = f (SEX, AGE, EDUC, AGREXP, EXTN, ACCREDIT, NOSO, INFOWTH, LVSTK, LANDSIZE, DEPRATIO, DIST) (6.2)

Where:

SEX	=	Sex of the head of the household.
AGE	=	Age of the head of the household in years.
EDUC	=	Education level of the head of the household in years.
AGREXP	=	Farming experience of the head of the household in years.
EXTN	=	Frequency of the household's contact with extension agents in a year.
ACCREDIT	=	Households access to credit; a dummy which takes the value of 1 if has access and zero otherwise.
NOSO	=	Number of social organizations that the household is a member.
INFOWTH	=	Regular access to weather information; a dummy which takes the value of 1 if has access and zero otherwise.
LVSTK	=	Livestock capital in Tropical Livestock Unit (TLU).
LANDSIZE	=	The size of the household's land holding in hectares.
DEPRATIO	=	Dependency ratio of the household as captured by the fraction between economically inactive population and economically active population in the household. Some studies include household size in general (Mano and Nhemachena, 2006, Deressa et al., 2008) with the belief that bigger households could allot larger labor force for adaptation activities. However, household size partly includes economically inactive population and labor could be diverted to attend children and the old and household chores, and this could be better reflected by dependency ratio.
DIST	=	Distance to the nearest market in hours.

6.2. Empirical Model

From the aforementioned discussion, we see that adaptation is a dichotomous variable that measures whether or not the households have carried out an adaptive strategy as a response to climate variability and change. In such setting where the dependent outcome is a dichotomous, binary choice models have been commonly and extensively used in empirical research. Therefore, logit model was adopted to analyze the adaptive behavior of the households.

As described by Gujarati (2004), the probability of obtaining a behavior can be specified as:

$$P_i = E(Y = 1|X_i) = \frac{1}{1+e^{-(\beta_0 + \beta_j X_i)}} \quad (6.3)$$

Substituting $(\beta_0 + \beta_j X_i)$ by Z_i, the equation would be:

$$P_i = \frac{1}{1+e^{-Z_i}} = \frac{e^{Z_i}}{1+e^{Z_i}} \quad (6.4)$$

Where:

$P_i = E(Y = 1)$ is the probability that the household performs adaptation strategy.

Z_i is a set of explanatory variables of the ith household (in this case it represents the variables specified above in subsections 6.1.1 and 6.1.2).

$\beta_0 and\ \beta_j$ are the parameters to be estimated.

If P_i, the probability of adapting, is given by equation 6.4, the probability of not adapting, is expressed as:

$$1 - P_i = \frac{1}{1+e^{Z_i}} \quad (6.5)$$

From this, the odds ratio in favor of adaptation could thus be:

$$\frac{P_i}{1-P_i} = \frac{e^{Z_i}/1+e^{Z_i}}{1/1+e^{Z_i}} = e^{Z_i} \quad (6.6)$$

Since, logit model uses logarithmic transformation to assume linearity of the outcome variable on the explanatory variables, the logit model could thus be expressed as:

$$\ln\left(\frac{P_i}{1-P_i}\right) = Z_i = \beta_0 + \beta_j X_i \quad (6.7)$$

6.3. Assumptions and Model Performance in Logistic Regression

Coefficients in logistic regression are estimated using maximum likelihood estimator. Unlike OLS, logit model does not assume all of the classical linear assumptions. It rather assumes some basic assumptions that help to make valid statistical inferences. The discussion here under briefly addresses the assumptions. The discussion also reviews briefly about model evaluation as logistic regression lacks consensus on evaluation methods and diverts from the conventional methods used in linear regression models.

Correct Model Specification

Correct specification of the model is one of the important assumptions in logit model. Particularly in the words of Menard (2001), correct functional form, inclusion of relevant variables and exclusion of irrelevant variables are indispensable for correctly specified model. Like the OLS, the logit model has to live up to the expectation of linearity in its functional form. Thus, the logit of the dependent variable as an outcome is assumed to be linearly related to the independent variables.

Model specification in logistic regression can be tested with **linktest** command in STATA program. The logic behind **linktest** is that there should not be any statistically significant additional predictor if the model is correctly specified. The **linktest** basically uses two predicted values - the linear predicted value by the model (_hat) and the square of linearly predicted value (_hatsq) as explanatory variables to reconstruct the model(http://www.ats.ucla.edu/stat/stata/webbooks/logistic/chapter3/statalog3.htm, accessed 28/07/2012). If the specified model does not need any additional variable and retains the correct functional form, the linear predicted value (_hat) has to be statistically significant, since it represents the model under consideration. Conversely, linear predicted value squared (_hatsq) should not be statistically significant or has much predictive power. If _hatsq becomes significant, this will be an indication that either there is omission of relevant variable(s) or the function is misspecified (Ibid).

Absence of Prefect Multicollinearity

Discussion on multicollinearity is already delivered in chapter 5. Here the brief discussion highlights only the technique of testing. Multicollinearity can be detected straightforwardly by VIF/Tolerance statistics. But, such tests are not issued in logistic regression due to its functional form. According to Menard (2001), the functional form of the model is, however, not important in detecting multicollinearity. Rather the relationship between the independent variables is typically an important characteristic that draws our attention when we deal with multicollinearity. Therefore, he proposes to run a linear regression model with the same variables used in logistic regression and compute for VIF/Tolerance test. This recommendation has thus been followed in this study.

Model Performance (Over all Model Evaluation and Goodness-of-fit)

According to Peng et al (2002), an assessment of the performance of a logistic regression model should cover the overall model evaluation, statistical test of the individual predictors, goodness-of-fit statistics and validation of predicted probabilities. Evaluating the overall performance of the model involves assessing the full model (a model containing predictors) against the reduced model (which serves as a baseline model) that only contains the intercept. An improvement of the full model over the reduced model is an indication for the better fit of the logistic model and a significant chi-square statistics is a sign for that. To this end, Likelihood ratio[79] test, which is an equivalent to the overall F-test in linear regression, is customarily issued (Gujarati, 2004; Field & Miles, 2010). The test computes the chi-square distribution from the deviance of the log likelihood of the two models and tests the null hypothesis that the predictors have no effect on the probability of the outcome. On the other hand, the significance of individual explanatory variables is tested using Wald test, which is easily obtained by dividing the coefficient by the standard error of the respective explanatory variables.

As regard to goodness-of-fit, the conventional measure in linear regression models is R-squared, which measures the proportion of variance in the dependent variable explained by a set of independent variables. However, such measure in logistic regression is not meaningful as the dependent variable takes only two values and thus the binary model does not form a straight line (Maddala, 1992; Gujarati, 2004). Because of this, several methods have been proposed. These methods mainly evaluate the model either by its precision to fit the estimated probabilities to observed responses or through the model's accuracy to predict the observed responses (Maddala, 1983). Nevertheless, there is no consensus on a single best method, and the methods may give different numeric figures (Pampel, 2000). Hence, Pampel (2000) advises researchers to use these methods carefully without greatly sticking to the exact figures. The most widely known coefficients of determination (Pseudo R^2) in logistic regression that mimic the linear regression model include McFadden R-squared[80], Cox and Snell R-squared[81] and

---

[79] The likelihood ratio is computed by the formula: $\chi^2 = 2\left[\mathrm{LL}_{\mathrm{Full}} - \mathrm{LL}_{\mathrm{Intercept}}\right]$; with $df = k_{\mathrm{Full}} - k_{\mathrm{Intercept}}$, where $\mathrm{LL}_{\mathrm{Full}}$ represents the log likelihood of the full model, and $\mathrm{LL}_{\mathrm{Intercept}}$ stands for the log likelihood of a model containing only the intercept (constant) (Field, 2005).

[80] As specified in Maddala (1992), McFadden R^2 is calculated by the formula: $R^2 = 1 - \frac{LL_{Full}}{LL_{Intercept}}$.

[81] Cox and Snell R^2 is given by $R^2_{CS} = 1 - \exp\left[-\frac{2}{n}\left[LL(Full) - LL\ (Intecept)\right]\right]$; where n symbolizes the sample size (Tabachnick & Fidell, 2007).

Nagelkerke R-squared[82]. Such numerous R-squares usually fuel controversies as to which method to use.

Another important measure of fitness, which is more often used, is Hosmer and Lemeshow[83] test. This test attempts to compare real world situation with predictions by measuring to what extent the actual (observed) events correspond to the predict events. The chi-square values are used to test the null hypothesis that there is no significant difference between the observed events and predict events, which in this case, insignificant results indicating the good fit of the model. In this study, all the aforesaid measures are reported and when further explanations are deemed necessary, Hosmer and Lemeshow test is used as this test seems attractive by verifying the predicted probabilities against the reality (observed probabilities).

Apart from model evaluation through the aforementioned methods, logistic regression model is further evaluated for its predictive power. Such evaluation mainly looks into the degree of conformity of predicted probabilities with that of observed outcomes (Peng et al., 2002). Classification table documents the validity of predicted probabilities by showing the proportion of correctly and incorrectly classified predictions. One aspect in classification table is what Maddala (1992) calls as count R-squared[84]and this statistics measures the overall proportion of correct predictions and could also be used to assess the improvement of the model's predictive power over the intercept model (prediction by chance). Along with this, classification table presents other important tests such as sensitivity and specificity statistics. While sensitivity measures the percentage of correctly classified events (in our case the proportion of households who have adapted), specificity measures the proportion of correctly classified nonevents (those who have not adapted) (Peng et al., 2002).

---

[82] Since Cox and Snell R^2 never reaches the maximum value of 1 even the model predicts perfectly, Nagelkerke R-squared amends Cox and Snell R^2 and is expressed as: $R_N^2 = \frac{R_{CS}^2}{R_{MAX}^2}$; where $R_{MAX}^2 = 1 - \exp[2(n^{-1})LL(Intercept)]$ (Tabachnick & Fidell, 2007).

[83] The test is performed by arranging predicted probabilities in ascending order and then sub-grouping into deciles of g groups. From this grouping, Pearson chi-square statistics is calculated from the 2 x g groups of observed and expected frequencies, using the formula: $x_{HL}^2 = \sum_{i=1}^{g} \frac{(O_i - N_i\bar{\pi}_i)^2}{N_i\bar{\pi}_i(1-\bar{\pi}_i)}$; where N_i is the total frequency of subjects in i^{th} the group, O_i is the total frequency of event outcomes in the ith group, and $\bar{\pi}_i$ is the average estimated predicted probability of an event outcome for the ith group (http://support.sas.com/documentation/cdl/en/statug/63033/HTML/default/viewer.htm#statug_logistic_sect039.htm, accessed 02/08/2012; Hosmer & Lemeshow, 2000).

[84] Count R^2 is given by: $= \frac{\text{number of correct predictions}}{\text{total number of observation}}$; where predicted probabilities greater than 0.5 are classified as 1; and those values less than 0.5 are classified as 0. Maddala (1992) treats count R^2along with other forms of R^2 to evaluate the goodness-of-fit of a logistic model.

6.4. Descriptive Analysis

In this section, adaptive strategies implemented by farmers and barriers that impede farmers' adaptive efforts are discussed briefly.

6.4.1. Farmers' Adaptive Strategies

It is already discussed in chapter 4 that nearly all of the households have perceived changes in the climate. Besides, focus group discussions offered an in-depth analysis of the behavior of the local climate by laying out several of its manifestations that appear to be compromising farmers' livelihoods, particularly of crop farming. Consequently, some farmers are attempting to contend the threats posed by the turbulent climate by implementing various strategies and others seem to do nothing special about it.

As presented in Table 6.1 below, the majority of households (65.2 percent) have employed adaptive strategies in contrast to 34.4 percent who have done nothing. Among the strategies employed by the households, changing of planting dates and using different crop varieties are found to be the most used ones followed by increased use of fertilizer and planting trees. On the other end of the spectrum, increased use of irrigation, diversification into non-farming activities and water harvesting are found to be the least used strategies. But, it has to be noted that most households employed multiple strategies rather than limiting with one specific strategy. This may show the probability that no one adaptive strategy is adequate to withstand climate predicaments that farmers are facing.

Table 6.1. Adaptive strategies carried out by the households

<table>
<tr><th>No.</th><th>Adaptation Strategy</th><th>No.</th><th>Percentage</th></tr>
<tr><td>1</td><td>Using different crop varieties</td><td>111</td><td rowspan="9">65.2</td></tr>
<tr><td>2</td><td>Changing planting dates</td><td>128</td></tr>
<tr><td>3</td><td>Adopting drought resistant crops</td><td>52</td></tr>
<tr><td>4</td><td>Increased use of soil and water conservation techniques</td><td>55</td></tr>
<tr><td>5</td><td>Diversification into non-farming activities</td><td>19</td></tr>
<tr><td>6</td><td>Water harvesting</td><td>11</td></tr>
<tr><td>7</td><td>Planting trees</td><td>75</td></tr>
<tr><td>8</td><td>Increased use of irrigation</td><td>29</td></tr>
<tr><td>9</td><td>Increased use of fertilizer</td><td>77</td></tr>
<tr><td>10</td><td>Done nothing</td><td>86</td><td>34.4</td></tr>
</table>

Source: Own field research data

While this might be true, some households may even apply the strategies not really as a response to climate stimuli but for other reasons such as profit or increasing productivity. In connection to this, it has already been discussed and argued in chapter 2 that one of the problems in adaptation studies is to identify as to whether adaptation strategies that has been performed by agents are climate driven or not. An attempt is made in this study to address this concern by following the methodological approach outlined in section 3.6 of chapter 3. Based on this methodology, nearly 11 percent of the households (27 in number) said that they did not perform any adaptive strategy as a response to climate variability and change while they have already carried out one or some of the adaptive strategies listed in Table 6.1.

Beyond the methodological remedy it offers to identify whether farmer's adaptive behavior is climate or non-climate driven, the result may lead us to raise some basic issues, and even in some cases to new openings for further research. First, it shows the likelihood of implicit learning of adaptation where farmers might not even notice that they are adapting to climate variability and change while actually implementing some of the adaptation strategies. From this view point, it could be possible to infer that adaptation itself is partly an elusive subject matter where our understanding still remains rudimentary, like that of climate change. Second, farmers who are seemingly carrying out adaptive strategies but who claim not carrying out any response action to climate stimuli (11 percent of the households stated above) might not be deliberately carrying out adaptation strategies as a response to climate effects. Nevertheless, the adaptive strategies that these farmers carry out (for whatever reason) will continue to act against the effects of climate variability and change. In this case, how do these farmers adaptive behavior be considered in the face of climate variability and change, i.e. should we consider them as adapters or non-adapters? Answering this question is very important in any attempt to understand adaptive behavior specifically targeting climate variability and change. As a matter of fact, addressing these issues certainly requires further research and follow-up studies. In the context of this research, since great care was taken to confirm households' responses by repeated process[85] and nearly all of the respondents have been well aware[86] of the changes in the local climate, it is more likely that these households' adaptive behavior might not be climate driven.

6.4.2. Barriers to Adaptation

Before passing into econometric analysis of the factors that facilitate adaptation, a particular attention is paid in this subsection to assess the factors that hamper

[85] The detail of the procedure is described in section 3.6 of chapter 3. When drawing their answers to the questions pertaining to adaptation, the respondents were asked once again to confirm their answers in cases where there was discrepancy between implementation of the strategies and adaptive behavior.

[86] Focus group discussions and the responses that the households gave in the questionnaire showed that nearly all of them believed that the local climate has changed.

adaptation. The assessment is carried out based on information obtained from farmers themselves through the questionnaire. As depicted in Figure 6.2 below, lack of financial capacity, lack of awareness and technical knowledge, and lack of suitable crops and improved seed are found to be the first three major problems (in terms of frequency counts) mentioned by farmers. In fact, adaptation requires financial resources to purchase inputs and lacking these resources implies less to spend on adaptation. Based on the characteristics of subsistence farming, where much of agricultural production is consumed by the household and there are fewer sources of income, it might not be as such surprising if farmers point out lack of finance as the most frequent problem that they encounter in adaptation efforts.

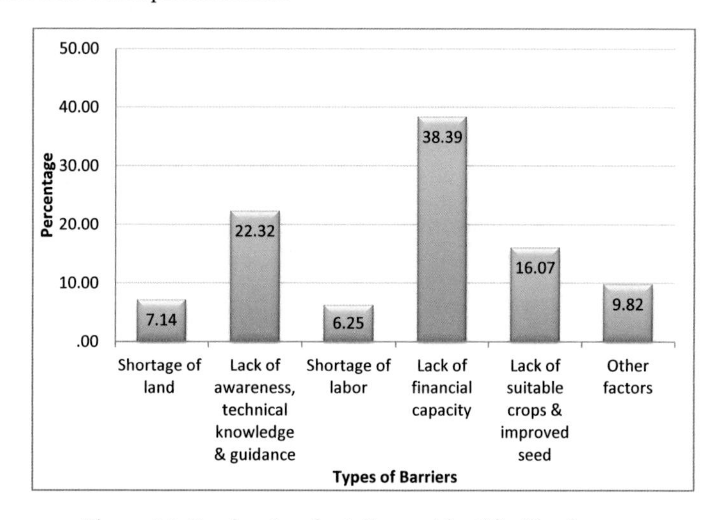

Figure 6.2. Barriers to adaptation as identified by farmers

Source: Own field research data

The other problem is lack of awareness and technical knowledge on adaptation methods. Lack of these factors could be the reflection of lack of information outlet, technical personnel and guidance on adaptation options, how to adapt and what adaptation strategies could be effective. In fact, the government has deployed three agricultural development agents (as stated in chapter 3) in each village to provide farmers with extension services. Even if farmers acknowledge the importance of the agents in providing information on agricultural technologies and natural resources management, there seems however a clash between farmers and the agents, as noted at the time of the field work. Unlike the consensus between the farmers and these agents as discussed in chapter 4 in characterizing the happenings in the local climate, they entrain major differences in the topic under discussion here.

As a matter of fact each of the agricultural development agents is either specialized in crop, livestock or natural resources management so as to deliver extension services to farmers as part and parcel of the government's policy to boost agricultural production and productivity. Towards this end, the agents dictate farmers to implement agricultural packages and prescriptions pushed down from the top authorities. These agents assert that they educate and advise farmers but also complain about the farmers' reluctance in putting the advices into effect (practice). Particularly, the complaint was clearly observed in two of the Weredas. On the other hand, farmers blame the agents having only equipped with words that lack tangible practical skills and support. In fact, the agents seem to be sandwiched between farmers' reluctance and authorities' pressure from the respective Weredas. On the one hand, farmers demand timely supply and affordable price of inputs, and on the other hand, authorities push the agents to enforce farmers implement agricultural packages. The agents feel that beyond the reluctance of farmers, lack of timely supply and higher price of agricultural inputs have been hampering farmers' effort in implementing adaptation and productive technologies that the agents recommend. That is why farmers' responses indicate lack of improved seeds and suitable crops (drought resistant) as the third most cited obstacles in adaptation.

In the first place, policy-driven adaptation to climate change and variability is yet an emerging issue and the agents might lack theoretical and practical skills in adaptation science. On top of this, introducing new forms of adaptation could be difficult given the country's low levels of agricultural research and development (R & D). Consequently, most of the adaptation methods applied by farmers are traditional ones and the agents are mostly limited to advise farmers to use these strategies through diversification & intensification, i.e., from relying on one specific method into adopting a number of methods; from the usual use of a method into increased use of that method (for instance, increased application of fertilizer, increased use of soil and water conservation techniques and increased use of irrigation).

However, there are a number of problems in implementing these advices. One of the problems, as indicated above, is lack of timely supply of inputs (mainly seed and fertilizer). As per the farmers' description, supply of inputs is a serious problem that they face almost in each season. Provision of inputs is usually promised by the authorities (through the agents) but mostly the delivery fails to be on time. Even in some cases, counterfeit seeds are provided by merchants and a number of farmers have already lost due to failed harvests. These situations may in fact create lack of trust between the agents and the farmers. Still some farmers equally view the agents as political agents of the ruling party who are deployed in villages to control and enforce them to do what the government wants them to do. In this regard, two agents from one village acknowledged that a considerable amount of their time is spent engaging in political activities. Such political engagement of the agents might have repercussion in farmers' reluctance beyond the natural unwillingness of human beings to embrace new technological changes (if any) easily. Further, time that could have spent on assisting farmers might be spent on political activities.

Other problems listed by farmers also include shortage of land and labor. Land is obviously crucial to implement and sometimes test the effectiveness of adaptive strategies before scaling up in the farms. Farmers in Ethiopia are largely confined to small fractions of land holdings (an average of 1.22 hectares per household (CSA, 2012)), and thus land size is not only a constraint in adaptation but also regarded as a poverty trap among agricultural households in rural Africa (Jayne et al., 2003). Other than these, factors such as population pressure (could be related with ever growing shortage of per capita land), and frequent occurrence of drought (could be related with the farmers' sense that adaptation methods have been destructed due to repeated drought occurrence & thus farmers might see frequent drought as obstacle by itself and beyond the coping capacity of the exiting adaptive strategies) have also been cited as barriers to adaptation.

In a nutshell, these all problems could be linked with poverty and lack of agricultural research and development (R & D) that helps to design and introduce new adaptive strategies that could be effective in the local context.

6.5. Regression Analysis and Discussion

So far, adaptive strategies and barriers to adaptation from the perspectives of farmers have been discussed. Even though farmers list some basic barriers to adaptation, some farmers still manage to implement some strategies while others fail to do so. What factors are responsible for this? To answer this question, it is time now to turn into regression analysis. Before going directly to the analysis, let us first resolve the issue of the households who have already implemented adaptive strategies for non-climate reasons. How should we consider these households in our regression analysis, i.e. as adapters or non-adapters? Two rivalry ideas as - pro-adapters and pro-non-adaptors - could be raised here. From pro-adapters view point, one can argue that as far as they implement adaptive strategies that can possibly offset the consequences of climate predicaments, they should be considered as adapters and the regression analysis should include them as adapters. The concern, as per this argument, is not merely on adaptive behavior as perceived by the households but more importantly on the actual implementation of adaptive strategies regardless of the reasons behind the implemented adaptive strategies.

On the other hand, one can also counter argue pro-adapter's assumption. Inclusion of these households as adaptors certainly carries a problem that while their adaptive strategy is non-climate-induced, it assumes as climate driven and hence impossible to discern the specific factors that influence adaptive behavior in the context of climate variability and change. Rather they should be regarded as non-adopters as reported by themselves. Still this approach carries a problem that while they already share the same strategies with those households who believed their behavioral responses as climate-induced, it ignores the importance of the adaptive strategies in reducing climate risks. Further it might have the risk of undermining implicitly learned adaptive strategies that goes unnoticed by farmers as climate-driven.

Given such contrasting arguments and our specific interest on climate-induced adaptive behavior, it seem more plausible and less risky to drop these households than facing the problem by including the households either as adapters or non-adopters. Hence, the regression analysis was performed by excluding these households and the results' discussion here under follows this approach. Nevertheless, for comparison and to see the effects of the above two arguments in our analysis, regression was also run under each of the assumptions and the results are reported in Appendix 10 along with brief explanation.

6.5.1. Impact of Perception Factors on Adaptive Behavior

Model evaluation test results listed at the bottom of Table 6.2 show that the model fulfills the assumption of linearity. _hat, which denotes the linear predicted value, is significant with the corresponding insignificance value of the _hatsq. This indicates that the relevant variables have been included in our model and the functional form is correct. The model is also free from multicollinearity as the VIF for all of the explanatory variables is much less than the threshold level of 10.

The significance of the overall model test (χ^2= 80.34, $P < 0.000$ with $df = 18$) indicates for the need to examine closely the predictors which turn to explain adaptive behavior better than the intercept only model. The independent variables explain between 28.1 to 42.3 percent of the variation in adaptive behavior. The Hosmer and Lemeshow test yielding an insignificant chi-square value of 12.08 ($P > 1$) also suggests that predicted probabilities of adaptive behavior fitted well with the actual adaptive behavior. Finally, the model correctly classifies 91.24 percent of those who have adapted and 61.25 percent of those who did not apply any adaptation strategy. The model's overall success rate of prediction is 80.2 percent which is by far larger than 63 percent of prediction by chance. This means that including the explanatory variables in the model reduces the classification error from 0.37 to nearly 0.20, indicating a drop by 46.2 percent. This shows that our model considerably improves the accuracy rate of the prediction.

Table 6.2. Logistic regression results with perceptual variables

Dependent Variable: Adaptive behavior (1= adapted; 0 = did not adapt)

Variables	Coef.	Std. Err.	z	P>|z|	Odds ratio
PV					
3	1.607	0.803	2.000	0.045**	4.989
4	2.532	0.797	3.180	0.001***	12.579
5	2.711	1.396	1.940	0.052*	15.040
PS					
3	-0.882	0.835	-1.060	0.291	0.414
4	0.335	0.758	0.440	0.658	1.398
5	1.679	1.120	1.500	0.134	5.360
AE					
2	0.615	1.198	0.510	0.608	1.849
3	1.371	1.298	1.060	0.291	3.940
4	1.731	1.245	1.390	0.164	5.649
5	1.219	1.303	0.940	0.349	3.384
PC					
2	-0.310	1.015	-0.310	0.760	0.734
3	0.766	1.005	0.760	0.446	2.150
4	2.030	1.030	1.970	0.049**	7.613
5	2.436	1.355	1.800	0.072*	11.423
SE					
2	1.519	1.276	1.190	0.234	4.567
3	-0.036	1.516	-0.020	0.981	0.965
4	0.688	1.511	0.460	0.649	1.990
5	1.127	1.676	0.670	0.502	3.085
Constant	-4.616	1.475	-3.130	0.002***	0.010
Linearity test					
_hat	1.048	0.175	5.990	0.000***	
_hatsq	-0.048	0.082	-0.580	0.559	
Overall model evaluation			χ^2	*df*	*p*
Likelihood ratio test			80.34	18	0.000*
Goodness-of-fit					
Hosmer & Lemeshow test			12.08	8	0.148
McFadden R^2			Cox & Snell R^2		Nagelkerke R^2
0.281			0.309		0.423
Validation of predicted probabilities			Correctly classified	Sensitivity	Specificity
(for details see Appendix 8)			80.18%	91.24%	61.25%
Multicollinearity: Mean VIF 2.10, the detail is shown in Appendix 8					
NB: *** significant at 1%, ** significant at 5%, * significant at 10%					

Source: Own field research data

As regard to individual predictors, Wald statistics shows that perceived vulnerability and perceived cost are found to be significant predictors. Whereas, the other explanatory variables i.e., perceived severity, adaptive efficacy and self efficacy, are statistically insignificant. Despite this, these three variables generally show that when a farmer develops higher levels of perceived severity, adaptive efficacy and self efficacy, the propensity to adapt increases. Households who perceived that their crop farming is extremely and highly vulnerable are more likely (with odds of 15 and 13 times respectively) to carry out adaptive strategies than those who feel less vulnerable. Similarly, those households who are confident of their resource base (financial capacity) are about 11 times likely to adapt than those who feel unconfident of their resource base. These results might not be as such surprising since increased levels of perceived vulnerability amplifies risk perception and increased confidence to afford the cost of adaptation enhances the competence to carry out adaptation strategies.

Further, the results could implicate two things. The first is that the concern to climate variability and change in the form of perceived vulnerability seems to be more important for adaptation to occur than signal detection (simple perception) of climate change. It is discussed in chapter 2 that a number of micro-level studies assumed adaptation as a two-step process (signal detection that the climate is changing[87] and then adaptation). In this premise, deficiency of signal detection is ultimately taken as a psychological barrier to adaptation and socioeconomic factors as objective barriers in the existence of signal detection. However, it is argued in this study against this premise that signal detection by itself cannot lead to adaptation and thus revealing a missing link between adaptation and signal detection. The results here clearly support this argument that while the households were able to detect the signals, it is those households who decoded this signal detection into higher levels of vulnerability that more likely tended to adapt.

We have seen in chapter 4 that nearly all the sample households have perceived changes in the climate. Even though nearly all of the households perceived the changes, not all of them have undertaken adaptation strategies. Now we further see that nearly all have perceived the changes but differ in their levels of perceived vulnerability to the perceived changes. We also found that households who have developed the sense of higher vulnerability tend to adapt more likely than those with the sense of lesser vulnerability. Here the link to adaptation clearly turns to be perceived vulnerability and hence nullifying the two-step process of adaptation. In other words, it means that signal detection clearly fails to carry a discriminatory power over adaptive behavior and the results rather underline cognitive process of adaptation requiring a condition more than signal detection. In connection to this, Risbey et al., (1999) describe adaptation as a process entailing several steps: signal detection (recognizing the signal and the noise), evaluation of the signal, decision and response (adaptive behavior) and feedback

[87]As captured by increasing or decreasing trend of rainfall and temperature.

(monitor the outcome). Evaluation of the signal involves interpretation of the signal and assessment of probable consequences by agents (individuals, agencies etc). It is based on the outcome of the evaluation process that behavioral responses are sought. Therefore, signal detection per se cannot be a psychological barrier to adaptation. Particularly this is true in the current era where agents are forced either to adapt to *climate change discourse* or *climate change disruption*.

Given such state of affairs, signal detection could be easily achieved by the discourse and/or disruption (experiential learning). *Climate change discourse* here implies to ongoing socio-political and scientific dialogue on climate change at global stage, where as *climate change disruption* refers to real changes in the climate that are already happening and experienced by agents (individuals, communities and societies) at local levels. If we are talking about adaptation to the discourse, it may be true that signal detection as awareness creation could be a precondition for adaptation process. On the other hand, if we are talking about adaptation to climate change disruptions, signal detection per se could only have little role as people would have ubiquitous awareness through experiential learning. In this later condition, adaptation process thus becomes the function of risk and adaptive capacity of the agents, which we tried to capture risk in the form of perceived vulnerability and severity, and adaptive capacity in the form of perceived adaptive cost, efficacy and self efficacy. Interestingly, it is found that perceived vulnerability which attributes perceived risk, and perceived cost which attributes perceived adaptive capacity as the significant predictors of adaptive behavior from the psychological perspective.

The second point that the results implicate goes with the competence to carry out adaptation. While perception of vulnerability serves to arouse the households' motivation to adapt, the competence to carry out adaptation more likely rests on perceived costs of adaptation. As perceived vulnerability is a product of the evaluative process of the external factors (assessing climate variability and change), perceived cost involves the appraisal process of the internal factors of the households, mainly resource endowments. Therefore, our analysis of adaptive behavior from psychological perspective shows that resources are still central to adaptation among subsistence farmers. Nevertheless, perceived cost is an abstract as well as a composite factor of several socioeconomic and institutional resources. Identifying specific enabling socioeconomic and institutional resources would thus be an important step to give guidance for policy making based on the specific results obtained from the regression analysis. The next subsection addresses this specific issue.

6.5.2. Impact of SEI Factors on Adaptive Behavior

The results of the regression are presented in Table 6.3. Our model in this case also passed the tests of linearity and multicollinearity. The overall model with chi square value of 55.67 and a probability of $P < 0.000$ ($df = 18$) indicates that the set of explanatory variables have a significant effect on adaptive behavior. These explanatory variables explained between 19.5 to 30.9 percent of the variation in adaptive behavior. The insignificant Hosmer and Lemeshow's goodness-of-fit statistics supports the hypothesis that the model fits the data well. With regard to prediction success rate, the model correctly classified 81.75 percent of those who have adapted and 58.75 percent of those who did not. Its overall prediction success rate stands at 73.3 percent, which is better than prediction by intercept only model (63 percent). Taking the explanatory variables into account cuts the classification error by 27.8 percent (shrinking it from 0.37 to 0.267). Similar to the earlier model, our model based on socioeconomic and institutional variables improved the prediction success rate. However, some of the statistical values show that this model is relatively less accurate than the model based on psychological variables. Simple comparison of the R-squares and the rates of correct predictions reveal the differences. We will come later to this point.

Table 6.3. Logistic regression results with socioeconomic variables

Variables	Coef.	Std. Err.	z	P>|z|	Odds ratio
1.SEX	0.844	0.459	1.840	0.066*	2.326
AGE	0.006	0.030	0.200	0.839	1.006
EDUC	0.008	0.058	0.140	0.887	1.008
AGREXP	0.034	0.029	1.180	0.239	1.035
EXTN	0.026	0.018	1.430	0.153	1.026
1.ACCREDIT	0.885	0.390	2.270	0.023**	2.423
NOSO	-0.519	0.256	-2.030	0.043**	0.595
1.INFOWTH	0.833	0.366	2.280	0.023**	2.299
LVSTK	-0.035	0.098	-0.360	0.720	0.965
LANDSIZE	0.106	0.278	0.380	0.704	1.111
DEPRATIO	-0.045	0.135	-0.330	0.739	0.956
DIST	-0.186	0.255	-0.730	0.468	0.831
Constant	-2.061	1.044	-1.970	0.048**	0.127
Linearity test					
_hat	1.163	0.219	5.310	0.000***	
_hatsq	-.1419	0.118	-1.200	0.228	
Overall model evaluation		χ^2		*df*	*p*
Likelihood ratio test		55.67		12	0.000*
Goodness-of-fit					
Hosmer & Lemeshow test		13.22		8	0.1046
McFadden R^2		Cox & Snell R^2			Nagelkerke R^2
0.195		0.226			0.309
Validation of predicted probabilities (for details see Appendix 9)		Correctly classified		Sensitivity	Specificity
		73.27%		81.75%	58.75%
Multicollinearity: Mean VIF 1.980, the detail is shown in Appendix 9					
NB: *** significant at 1% level, ** significant at 5% level, * significant at 10% level					

Source: Own field research data

Among the explanatory variables, sex, access to credit, membership in social organizations and access to weather information are statistically significant in explaining adaptive behavior. Male-headed households are found to be more likely to implement adaptation strategy(ies) as compared to their female-headed counterparts (the odds of adapting by male-headed households is 2.3 times that of female-headed households). This result is consistent with other studies on related fields (Deressa et al., 2008; Nabikolo et al., 2012). The justification might go with rural women's less control over resources, technology, extension services and information which are considered to play an important role on adaptation process (Deressa et al., 2008; Meinzen-Dick et al., 2010; Nabikolo et al., 2012). On the contrary, Nhemachena & Hassan (2007) found female-headed households in South African region more likely engaging in adaptation actions to climate change. So, the impact of sex on adaptive behavior could be culture specific and might vary based on social structure (Nabikolo et al., 2012).

Access to credit is another important factor found to affect adaptive behavior positively. The probability of adaptation strongly increases as households get access to credit services with odds ratio of 2.423. This indicates that households who have access to credit are over 2 times more likely to adapt than those who do not have credit access. A number of studies also reported similar results (Maddison, 2007; Nhemachena & Hassan, 2007; Deressa et al., 2008; Gbetibouo, 2009; Nabikolo et al., 2012). It appears that access to credit relaxes financial constraints of the households to purchase inputs and meet transaction costs related with adaptation (Nhemachena & Hassan, 2007).

Likewise, access to weather information is found to be a strong predictor of adaptation. Getting weather information increases the odds ratio of adaptation by a factor of 2.3. This result corresponds with Deressa et al (2008) who reported an increasing likelihood of adaptive behavior as a response to access to climate information. Based on farmers' reports, Maddison (2007) also indicated lack of information about weather and climate being as one of the barriers to adaptation. In fact, provision of user based day-to-day weather information keeps farmers up-to-date and hence helps them decide on adaptation options. For instance, if farmers get timely seasonal information on weather conditions, they can easily decide whether there is a need for switching planting dates given credibility of the source of information.

Another result which seems to be surprising is membership in social organizations. The literature acknowledges the importance of social organizations (as a form of social capital) in enhancing adaptive capacities of individuals (Adger, 2001; Adger, 2003; Halsnæs & Verhagen, 2007; Pelling, 2011). Through networks and social interactions with in organizations, members can have more access to resources such as information, risk management skills and other transferable knowledge which could be used for decision making in adaptation efforts. Despite this, it is paradoxical to notice membership in social organizations affecting adaptive behavior negatively and significantly. As seen in the table above (Table 6.3), increasing membership in social organizations by one unit decreases the probability of adaptation by 40.5 percent. It might be less clear why membership in social organizations negatively influence adaptive behavior. Nevertheless, some probable postulates could be made. Members might be spending a lot of time in non-productive activities, such as political networking and discussions, while engaging and participating in the organizations. The other plausible reason could be linked with the psychosocial impacts of climate variability and change discussed in section 4.5 of chapter 4. It was learnt from focus group discussions that social fabrics established for reciprocal altruism and helping the disadvantaged groups are not functioning well as farmers are forced to compete with the behavior of the rain (see the details in the section). Consequently, farmers can no longer be able to discharge the responsibilities vested on them being as members of the social fabrics. Such state of affairs in turn degrades the overall activities and operation of social organizations. This eventually might be reflected in a number of socioeconomic settings, and adaptation being one of them.

The other variables which include age, level of education, experience in agriculture, access to extension services and land size have positive effects on adaptation but are statistically insignificant. The positive sign with these variables is in accordance with expectation. In fact, age could have mixed effects - old farmers being more resistant to change and at the same time acquiring knowledge and experience on adaptation. Accordingly, age could have played either way (positive or negative). The relationship of the rest of the variables with adaptive behavior is obvious. Education and experience as sources of knowledge and skill, extension services as providers of climate information and adaptation options, and land as a reflection of wealth and extra space to practice adaptation options play positive roles in adaptation effort. On the other hand, livestock ownership, dependency ratio and distance to the nearest market have negative effects. The effect of dependency ratio and distance to the nearest market is clear. As dependency ratio increases, labor is more diverted into household chores and care provision to the economically inactive members. Similarly, as distance to market increases, access to inputs for adaptation is constrained by the distance. In consequence, adaptive behavior is affected negatively in both cases. Finally, the most probable reason for livestock's negative sign could be linked with farming preference or specialization of the households who are relatively wealthier with livestock capital. These households might be more inclined to animal husbandry and thus might be less active in adaptation activities targeted at crop farming.

Finally, let us conclude our discussion by viewing which of the models (psychological or SEI) seems to be more influential in explaining adaptive behavior. As seen in each of the above tables, psychological factors explain in between 28.1 to 42.3 percent of the variation in adaptive behavior. Whereas, socioeconomic and institutional factors do the same task in between 19.5 to 30.9 percent. Similarly, the model based on psychological factors correctly classifies 80.2 percent of the overall classification and the rate of this task by the model composed of SEI variables is 73.3 percent. Based on these figures, it seems that the model constituting psychological variables emerges to surpass the model with SEI variables, and may show the stronger influence of psychosocial variables on adaption decision. Similar result was also reported by Grothmann & Patt's (2005) study focusing on adaptive behavior against flood threats among residents in the banks of Rhine River, Cologne, Germany. In their study, they found out that socio-economic factors explained 3 - 35 percent of the variation and cognitive factors doing 26 - 45 percent of the variation.

This may be due to the following possible reasons. As far as threat of risk (e.g., perception of vulnerability) among individuals is high, there is a high probability that farmers try to respond in different ways not necessarily or solely based on socioeconomic variables. Another reason may be linked with perceived cost, which we discussed above. Perceived cost as one of the significant psychological variables is formed based on the appraisal process of SEI resources. This appraisal process involves several tangible and intangible resources that we would not be able to measure or capture in an empirical model. For instance, our model based on socioeconomic variables only includes some of these tangible and intangible resources. Even though, modeling with socioeconomic variables is more pragmatic than with that of psychological variables, still the explanatory power of the psychological variables seems to be more influential most probably due the reasons described here and hence showing the importance of psychological factors in understanding adaptive behavior.

Concluding Remark

In this chapter, going beyond the traditional approach of analyzing adaptation through tangible resources, adaptive behavior is framed and analyzed from psychological perspective using Protection Motivation Theory. Founding on a premise that PMT may provide a useful framework to understand individual adaptive behavior in the context of climate variability and change, this study demonstrated the explanatory power of the framework. In doing so, the role of resources has never been neglected. As such, adaptive behavior is also framed from socioeconomic perspectives.

Both of these approaches are found to be relevant and have yielded some insights. Although psychological factors seem to have an upper hand explanatory power over SEI factors, resources still emerge to be crucial for adaptation. This is already seen by two pieces of insights noted from the regression analysis. First, psychological factors themselves (through the perceptual appraisal process of costs of adaptation (perceived cost)) have implicated resources. Second, some important socioeconomic and institutional factors such as access to credit and climate information emerged to be influential for adaptation. Therefore, these two approaches are found to be complementary, rather than competing. While the psychological factors might be more important to measure the level of risk and motivation to adapt, resources might be more important to measure the level of competence of agents in adaptation process.

Among affluent societies where resources are not as such constrained, it might be plausible to find out psychological factors being more powerful than SEI factors in predicting adaptive behavior. Yet, psychological factors have a power to explain and refine our understanding of adaptive behavior among subsistence farmers where resources are meager and credible climate information is scarce. From this perspective, it is also possible to conclude that strengthening research on the psychological aspect of adaptation may help us better understand adaptive behavior and offer pragmatic instruments to influence behavior, and select effective adaptive strategies under the discretion of farmers who confront climate predicaments.

Chapter Seven

Conclusion and Outlook

7.1. Revisiting the Research Problem and the Gaps in the Literature

This dissertation has aspired to document and analyze local knowledge on climate variability and change, assess the impacts with special emphasis to quantify the economic impacts on crop production, and examine the conditions that govern adaptive behavior at micro-level. As the starting point, it reviewed the literature and unveiled the following gaps that have become pivotal in the overall process of this research.

The first gap noticed in the literature is related with little emphasis to local (indigenous) knowledge. Climate change discourse is entangled with uncertainties in the science, contradicting positions and debates among the believers and the skeptics, and disagreements on the scale and financing of responses. In such world of knowledge, where debates will continue to fuel until scientific advancement prevails with convincing theories and economic and political interests are alienated from the reality of climate change, it will probably take much time to bring world leaders and scientific community into consensus so as to effectively address local concerns and vulnerabilities of the poor in the developing world. On the other hand, even if world consensus is possibly reached, it is most likely that policies that are less sensitive to local climate and concerns would emanate from the climate change discourse and be pushed down to the localities due to the influential position of the discourse. So, in this study, it is argued that understanding local knowledge will have paramount importance in designing policies that are effective to address climate problems and vulnerabilities at local levels. It may also help to fill paucities in scientific data, particularly in the developing world where climate recordings are poor and sparse in spatial coverage. This task primarily depends on the analysis of local knowledge and its trustworthiness. To this end, an attempt is made in this study to understand detection mechanisms, attribution and impacts of climate variability and change through the lens of local knowledge and to verify this knowledge through objective knowledge (scientific) systems.

The second area of interest that the literature seems to be weak is in measuring the economic impact of climate variability and change on subsistence agriculture. The customary way of measurement involves the Ricardian approach, which tries to capture the impacts as reflections on land values. However, some of its assumptions that do not fit into the developing world coupled with other theoretical limitations make the approach very weak in the context of subsistence agriculture. Dwelling on agricultural production function, an alternative modeling that captures climate both as a source of total factor productivity and a random term (Solow residual) is proposed and operationalized.

The third point of paucity observed in the literature goes with little attention to psychological barriers in adaptation. There has been a range of research exercise to understand adaptive behavior of small scale farmers highly concentrating on socioeconomic and institutional factors. The role of SEI factors in facilitating adaptation is beyond doubt. However, this study has argued that perceived risks and subjective adaptive capacity of farmers could play an important role either by motivating or demotivating adaptive behavior even before SEI factors come to surface as real determinants. Hence, this study has highlighted the quest for understanding adaptive behavior not from SEI standpoint alone but also from psychological perspectives.

7.2. Summary of Major Research Findings

The research has pursued to attain its objective by employing mixed methods approach. The first theme of the research (Theme 1) analyzed local knowledge based on information gathered from farmers. Content analysis and statistical procedures were employed in analyzing the information collected.

As per the perception of farming households, precipitation has been declining over time with an increasing trend of the corresponding temperature. Statistical analysis of the trends of precipitation and temperature in the weather stations also revealed decreasing and increasing trends respectively in the two decades considered in the study (1990 to 2009). This result attests the conformity of farmers' perception with climate recordings of the weather stations.

Perception concerning the behavior of the climate variables, on the other hand, shows increasing fluctuations on the timing (onset and cessation) of rainfall, and increasing tendency on the odds of heavy rain, drought and flood occurrence. Days are also perceived warming from time to time. These descriptions of farmers are also found to be in line with experts' assessment. Such conformity of farmers' assessment both with meteorological data and experts' views implies the close match of farmers' subjective assessment with that of the empirical evidence.

Farmers also supported their detection claims through concrete instances. Very interestingly, warmness of days (increased temperature) is found to be accompanied by changes in the longevity of food and beverage fermentation and preservation period. As compared to the old days, farmers witnessed decreases in the length of fermentation period of food dough and beverage malt and the overall longevity of the final products. This also shows that as scientists attempt to illustrate climate change through ozone layer depletion, see level rise, polar ice sheet melting and so forth, local communities have also their own ways of meaningful detection mechanisms.

According to farmers, the causes behind the changes are population pressure, deforestation and God's will. This attribution is also supported by experts with the exception of God's will. Experts divert from the notion of God's will and instead link the causes with the global outlook of climate change, i.e. greenhouse gas emission. This shows that in scope the causes are localized and blended with culture when it comes to farmers, and goes beyond localities and embraces the thesis of climate change discourse when it comes to experts. Whereas in terms of the root causes, experts relate it directly to anthropogenic activities and farmers attribute the changes both to anthropogenic sources and God's will. The former root cause brings farmers in line with what is said largely in the scientific discourse. On the other hand, the later cause takes them back to religious beliefs, which is regarded as culture in the eyes of scientific discipline. Even though they believe in God's will and the changes are signs of punishment for disobeying Him, interestingly, they also put weight on man's infringement of nature's stewardship given to him by God. This *notion of stewardship*, as termed in this study, although seems to come literally under God's will, it is a cultural reflection of attributing the causes to human activities. Therefore, local knowledge directly or indirectly supports the idea of anthropogenic causes.

Following the changes in the local climate, farmers pronounce the following impacts in particular.

- *Impact on availability of water*: water flow in rivers and streams has been receding slowly over the years, and in some cases these water bodies which were used to flow the whole year tend to dry until the next rain season. Although climate change and climate variability have clear impacts on water resources, it is however found to be hard to link these accounts solely with local climate stimuli given increasing practices of small scale irrigation schemes in recent years throughout the country.
- *Impact on health*: new malaria invasion has been observed in some of the research sites. This could be a good example to assert how climate change induces malaria to the highlands, which were malaria free few decades ago.
- *Impact on crop production*: farmers also stressed that the timing of rainfall and its distribution over time are found to be serious concerns to crop production rather merely the amount. These features of rainfall have posed serious problems on crop production in general and extra impacts on female-headed households. Further impact on female-headed households is related with farmland renting behavior of these households. Due to lack of adult male labor, theses households rent their land and tenants often tend to prioritize their own land through timely farm management and decision making in cases of increased disturbances in the climate variables while attending the rented land on secondary bases (unparallel to the behavior of the rain).

- *Psychosocial impact*: due to unpredictability and increased fluctuations on the onset of rainfall, farmers get confused especially on the timing of planting, which eventually has repercussions on crop productivity. The other aspect of psychological impact is related to tearing the social fabrics (social capital) of assisting each other. There has been a long tradition among communities to help each other in land preparation, ploughing, sowing and reaping. In these activities, priority is given to disadvantaged groups (widows, disabled, old people without a working family member, etc) of a community where members come to work for these groups in each farming season. In the current era, this assistantship seems to be breaking as farmers are forced to compete with the timing of rainfall and its less predictable behavior, and thus inflicting them to confine in their farm lands for timely decision and farm practices. The assistance for the disadvantaged groups usually comes late after farmers attended their own farms, and ill-timed intervention has repercussions on productivity of the farmlands of these groups. Therefore, unlike the usual effects on socioeconomic factors, climate change and climate variability have also a dismantling effect on social capital (social fabrics).

Beyond documenting and analyzing local knowledge, the research has dealt with the impacts of climate variability and change and associated adaptation on crop production under Themes 2 and 3 of the research framework. Multiple regression analysis was employed in both cases. Accordingly, climate change in mid-21st century appears to have positive effects in the highlands of Ethiopia. However, climate variability as represented by rainfall's deviation from its long term mean and early cessation shows to carry negative and significant effects. This has also been supported by farmers who critically viewed the timing of rainfall and its distribution over time (climate variability) than merely the amount of rainfall (climate change). Therefore, climate variability as captured by some of its forms is an immediate and serious source of concern than climate change in the mid-21st century when it comes to the highlands of Ethiopia.

On the other hand, adaptation as a counteractive action has shown positive and significant effect on crop production, and thus it could help to reduce the negative impacts of climate variability and change. Upon realization of the positive outcomes of adaptation, the study finally analyzed the psychological and socioeconomic determinants of adaptive behavior. Among the psychological factors, perceived vulnerability and perceived costs are found to influence adaptive behavior significantly. This results show that advanced cognitive processes (perceived vulnerability/risk perception, and perceived costs/perceived competence) are the necessary conditions for adaptive behavior than mere signal detection. In addition, the psychological process of adaptation, i.e. perceived cost, still implicate adaptation requiring some enabling socioeconomic and institutional resources. In relation to this, sex, access to credit, membership in social organizations and access to weather information are found to be significant factors among the socioeconomic and institutional variables.

7.3. Theoretical Contributions and Policy Implications of the Study and Outlook

This study has some contributions to the frontiers of knowledge and policy. Theoretically, it fills the paucity in exploitation of local knowledge as an alternative epistemology to understand climate variability and change at local levels with vivid examples and eventually integrate it with scientific inquiry. It also enlightens the influence of psychosocial factors in influencing adaptation and contests the two step process of adaptation. Methodologically, it has demonstrated ways of analyzing local knowledge taking the analogy of scientific steps involved in the study of climate change and variability. It also proffers simple remedies to distinguish whether farmers' adaptation strategies are climate or non-climate driven. The following points recapitulate the synthesis, contributions, policy outputs and outlook of the research briefly.

7.3.1. Towards Local Discourse of Climate Variability and Change: Bridging the Gap in Climate Change Discourse

Even though climate change discourse is purely the product of scientific conclusion (Cobb, 2011), it still fails to bring consensus among the scientific world and able to convince climate skeptics. That is why Cobb (2011:3) interestingly stated that "scientific knowledge alone cannot move climate discourse beyond its current troubled state". It is in troubled state because it is mired in controversy, skepticism and inaction (Ibid). This simply implies the quest for multiple epistemological sources to enrich and reconcile the discourse so as to mollify the troubled state. If not, it further implies that as far as the discourse stays in troubled state, collective world response will also sustain in troubled state and hence leaving the vulnerable and the poor to further vulnerability. It is also apparent that political and economic interests embedded in the discourse make consensus and effective collective action out of reach more likely for many years to come.

In such standoff, it is prudent to look into local knowledge, as a source of knowledge, which would guide policies and decisions that take local contexts into account. In fact, some argue that local knowledge is key to better understand climate change, supplement scientific knowledge and incorporate into the global discourse, so that it will have valuable lessons to offer (Nyong et al, 2007; Macchi et al., 2008; Green & Raygorodetsky, 2010; Cobb, 2011; Raygorodetsky, 2011). Indeed, local knowledge is important in particularly to guide response actions and make informed decisions at the localities.

As demonstrated by this study, local knowledge appears to be one of the key epistemologies to understand climate variability and change in the localities. Despite its detachment from the global discourse, local knowledge has offered important insights accompanied by concrete examples. This knowledge which has its own epistemologies rooted in generations, observations, experience and culture is also found to be in compliance with the scientific knowledge in many of the instances which strengthens the credibility of the knowledge.

Despite differences in level and scale, local knowledge has detected what science strives to detect through various rigorous processes and sophisticated technological systems. It has offered meaningfully instances on the impacts (stated above under section 7.2) resembling with what the science endeavors to articulate. It has also provided both supporting and diverging ideas in one of the areas where the discourse is trapped, namely, in the area of attribution. One of the important lessons that emerged from this study goes with this area. Even though, attribution through local knowledge has a cultural footprint, it directly (through anthropogenic activities) or indirectly (through the *notion of stewardship*) supports the idea of anthropogenic causes.

Given the insights of local knowledge revealed in this study, as climate change discourse strives to guide policy decision at the global level, policy dialogue and decision-making at local and regional levels need to consult local discourse of climate change. While climate change discourse is partly embedded with disguised political and economic interests, local discourse is largely a byproduct of real socioeconomic happenings, environmental conditions and vulnerabilities in the localities. Thus, making it (local discourse) more close to reality and which in turn leads to the formulation of pragmatic actions of response. Local discourse of climate change and variability here refers to a set of collective knowledge system and societal dialogue on local climate that have grown from experience, observation and historical accounts. As could be seen in the table below, local discourse could help us to understand and analyze climate change and variability at local levels through the same main structural elements that the global discourse is built on, and based on that understanding, analyze impacts and develop appropriate policies.

Table 7.1. Comparing climate change discourse (global) and local discourse

Area of comparison	Climate Change Discourse	Local Discourse
Epistemological source	Scientific knowledge	Local knowledge
Detection	Historical meteorological and oceanographic recordings Atmospheric studies (ozone depletion, concentration of GHGs in the air) Ice sheet melting, see level rise	Perception, experience, historical changes in the behavior of the climate (trends, timing, amount, duration, extreme events) Concrete examples, such as changes in the longevity of fermentation and food preservation. Changes in water flows of rivers and streams etc.
Attribution	Anthropogenic activity + natural process Evolves purely from scientific exercise	Anthropogenic activity God and lack of stewardship Evolves mainly from experience and culture
Impact	Climate change is serious	Climate variability is serious
Response/Policy Approach	Top down approach Likely induces adaptation to the discourse	Grass roots response Mainly climate disruption induced adaptation

Source: Own synthesis from the overall research process

Although local discourse and global discourse differ in level, scope and emphasis, the two discourses could be used as platforms to share knowledge and benefit from each other. Mentioning some of the benefits would be worthwhile at this stage: As argued by Nyong et al (2007), Macchi et al (2008), Green & Raygorodetsky (2010), Cobb (2011) and Raygorodetsky, (2011), local discourse could be used as a source of experiential knowledge to supplement epistemological claims of the science through concrete examples. In this case, even if local knowledge is limited to the respective localities and in its scope, several cases from localities around the world could serve as living laboratories to study the climate and complement the discourse through experiential knowledge and tangible evidence. It is well understood that global climate change affects local climate and the effects are largely evidenced at local levels, and hence local discourse could help us to identify and understand such effects of global climate change.

From policy point of view, local discourse serves as a stepping stone to minimize the influence of *adaptation to climate change discourse*[88]. Although climate change discourse is a byproduct of scientific output (Cobb, 2011), it also entertains ingredients of social constructions of climate change as a derivative of the media, political and economic interest, and influence of powerful world players. Such construction more often emanates policies that reflect its fundamental nature (essence) and more likely promotes top-down policy responses. In other words, it leads more likely to *adaptation to the discourse* than *adaptation to climate disruptions*[89] in the localities. Historically, such policies tend to undermine local contexts and values. Eventually the implementation of these policies might be marred by local community's rejection or end up in destabilizing local response systems that enabled the co-existence of people with the environment. Consulting local discourse could thus be a remedial to avoid the evolvement of such policies by balancing the top-down approach and helping to frame local based response mechanisms.

Mutually, climate change discourse has a lesson to offer to local discourse. By engaging local communities to the wider discourse, it should work towards influencing culture driven misconceptions (e.g, climate change is God driven) and bring it aboard into reality, and hence gear the commitment of communities towards protecting the environment by enhancing *the notion of stewardship*.

Finally, research has to push further to document and understand local discourse, find elements that science might exploit, and work on the methods to enhance mutual benefit.

7.3.2. Towards the Very Form of Climate Critical to Subsistence Farming: Bridging the Gap in the Emphasis of Climate Concern

Another important lesson emerged from this study reflects the need to address the very climate concern at local levels. The impact of climate change has been widely discussed and articulated as a serious problem. Yet, little is known about its actual impacts. Contrasting to climate change (global) discourse, local discourse frames climate variability as a serious and an urgent problem (in the context of this study). It seems paradoxical that while climate variability is dismantling the livelihoods of local people (subsistence farmers and the poor), climate change discourse has been largely trapped in responding to future climate impacts, which by itself is insensitive to urgencies in localities.

---

[88] See the definition in section 6.5.1 of chapter 6.

[89] See the definition in section 6.5.1 of chapter 6.

Assisted by its powerful technocratic and political apparatus, the global discourse has drawn policy makers largely towards devising policy options as to how to respond to the impacts of climate change while paying little attention to current climate variability. However, both qualitative and quantitative results in this study show such policy approach to bear an old proverb which says "putting the cart before the horse". The positive effect of climate change on crop production in the mid-21st century in the highlands of Ethiopia and the adverse effects of current climate variability and farmers' serious concern to the effect of climate variability could exemplify this. Therefore, policies have to pay due attention to current climate problems faced by farming communities. More importantly, it is through addressing current climate problems that adaptive capacity for future uncertainties could be built on.

In addition, research should also venture to incorporate the impact of climate variability when assessing the impacts of climate change. Mostly, research activities have immersed to predict the impacts of climate change by assuming climate variability being constant (Reilly et al., 2001). On the other hand, climate change is expected to increase climate variability or induce changes in the basic elements that characterize climate variability (Sperling and Szekely, 2005; Kandji et al., 2006; Washington et al., 2006). In this case, it warrants that research has to strive to understand the impacts more importantly through the changes induced on climate variability. In fact, this will not be an easy task given the difficulty to quantify the effect of climate change on climate variability and it will be an uphill task to the scientific world.

7.3.3. Towards Integrated Approach to Adaptive Behavior: Bridging the Gap between Psycho-cognitive and Socioeconomic Approaches

At micro-level studies, adaptation is conceptualized to involve two steps: perceiving the changes and then decision to adapt. In this approach, adaptation is primarily linked with signal detection and it is after fulfilling this step that actual adaptation may take into effect. Nevertheless, this study fails to find a link between signal detection and adaptation; rather it finds a link between risk perception and adaptation. It therefore cautions against oversimplification to understand adaptation through simple signal detection. In this world where people have the awareness of climate change and variability directly or indirectly either from the global discourse or the local discourse, signal detection as a variable commanding adaptation seems to be weak and only serves as a superfluous psychological barrier to adaptation.

As discussed in chapter 2, the vast literature of social science research on risk also highlights risk perception as an important cognitive element in people's response. Similar to such fashion, this study has found higher levels of perception process (e.g. perceived vulnerability implicating risks, and perceived cost that of self competence) influencing adaptive behavior. Therefore, research has to pay attention to understand micro-level adaptation process from psychological perspectives as well as the customary socioeconomic perspectives, and work further to integrate the approaches. On the other hand, adaptation policy should work on disseminating information to help agents (farmers, individuals) appraise their vulnerability, probable severity of the impacts and efficacy level of adaptive measures as well as working towards building self efficacy of people. While the first three psychological processes involve communication (information) inputs, the last factor in fact requires developing skills and resource base of people that enable them implement adaptive strategies. For instance, in the context of Ethiopian highlands, the study warrants the government (policy makers) to work towards facilitating adaptation enabling conditions such as farmers' access to credit, climate information, and gender sensitive support schemes. Particularly, building trustable climate information system could help farmers to update and prepare themselves and facilitate the cognitive process of risk perception and informed adaptation decision. It also helps to minimize confusion among farmers about the dates of planting.

7.3.4. Towards Mitigation Actions at Local Levels: Bridging the Gap in Local Response

Adaptation is widely considered as a vital response mechanism to developing countries and has been highlighted an urgent priority for them (UNFCCC, 2007). Because of these, policy dialogue at international, regional and national levels has highly tuned towards issues of adaptation (funding and its mechanisms, capacity building and institutional set up, etc) when it comes to the responsibility of these countries in combating climate change. However, from the perspectives of farmers (as the results of this study show) population pressure, deforestation and lack of environmental protection (through *the notion of lack of stewardship*) are found to be the major causes for ongoing changes in the local climate. If local communities conceive these factors as the major problems (also science supports this), can adaptation be a viable response mechanism to solve these problems? Should policy continue to enforce farmers to adapt to the effects while paying little attention to the root causes that bring the effects in the first place?

In fact, adaptation enables farmers to adjust to the changes in the local climate caused by the aforementioned problems. Adaptation is still important to adjust to global and local environmental changes in general. However, overwhelming focus on adaptation could take us to the preventionist school of thought where communities undermine mitigative actions (addressing the root causes) and continue with their usual actions that exacerbate environmental degradation and instigate new vulnerabilities in the localities. Effective and sustainable solution could only be attained when adaptation is promoted to address already happening changes and the irreversible ones, and mitigation is brought along with adaptation to address the root causes. Therefore, a balance between adaptation and mitigation at local levels should have to be installed in any policy undertaking.

More specifically, as the case in Ethiopian context, local mitigation strategies should be designed to combat the root causes of changes at local levels. If one of the root problems is population growth, it should thus be time to reconsider population policy as a mitigative response in the context of climate change and variability. The same is in dealing with deforestation. Rather than preparing farmers to adapt and live with the effects of deforestation and environmental degradation in general, promote polices that enforce farmers to afforestation and environmental rehabilitation. Particularly agricultural development experts, who work with traditional societies where cultural values and practices are intact, need to motivate farmers' behavior towards environmental rehabilitation by integrating *the notion of stewardship* in the implementation of polices as stated in subsection 7.3.1 above. The *notion of stewardship* is meaningful for the communities and thus induces the sense of responsibility in them and could be used as an instrument to influence their behavior and draw their commitment to continuously engage in local mitigation actions.

7.3.5. Towards Addressing the Needs of the Disadvantaged Groups: Bridging the Gap in Social Networks and Institutional Arrangements

As summarized earlier, climate variability and change have special impacts on the disadvantaged groups of the community by posing extra impacts on their farms and indirectly through breaking social capital (values of assisting the needy). Apparently they continue to be more vulnerable due to ill-defined institutional arrangements and their distinctive socioeconomic characteristics that make them more exposed to adverse climatic events. Policies often address communities in general while overlooking disparities within communities. Therefore, policy has to ensure that the needs and specific vulnerabilities of disadvantaged groups within farming communities are addressed in particular through different mechanisms. For instance, in the case of Ethiopia, establish legal enforcement and land renting institutional arrangements that oblige tenants to manage the rented farms on time equally with that of their own. In addition, support traditional social networks to sustain societal values that have made social support and responsibility available for the disadvantaged groups.

References

Aberra, K. (2011). The impact of climate variability on crop production in Ethiopia: Which crop is more vulnerable to rainfall variability? *Paper presented at the 9th International Conference of EEA/EEPRY*, Addis Ababa, Ethiopia. Paper presented at the 9th International Conference of EEA/EEPRY Addis Ababa, Ethiopia.

Adams, R. M., Hurd, B. H., Lenhart, S. & Leary, N. (1998). Effects of global climate change on agriculture: an interpretative review. *Climate Res*, Vol. 11:19-30.

Adesina, A. A. & Baidu_Forson, J. (1995). Farmers' perceptions and adoption of new agricultural technology: Evidence from analysis in Burkina Faso and Guinea, West Africa. *Agricultural Economics* 13, 1-9.

Adger, W. N. (2001). Social capital and climate change. *Tyndall Centre Working Paper* No. 8, Tyndall Centre for Climate Change Research and CSERGE, School of Environmental Sciences, University of East Anglia, Norwich.

Adger, W. N. (2003). Social capital, collective action, and adaptation to climate change. *Economic Geography*, 79 (4): 387-404.

Adger, W.N. & Kelly, P.M (1999). Social vulnerability to climate change and the architecture of entitlements. *Mitigation and Adaptation Strategies for Global Change* 4: 253-266.

Adkins, L.C. (2009). *Using gretl for principles of econometrics*, (3rd ed.). Version 1.21. Oklahoma State University.

Ali, A., Mushtaq, K., Ashfaq, M., & Abedullah (2008). Total Factor Productivity (TFP) growth of agriculture in Pakistan: Trends in different time horizons. *Pak. J. Agri. Sci.*, Vol. 45(4).

Armitage, C. J. & Mark, C. (2000). Social cognition models and health behavior: A structured review. *Psychology & Health*, Vol. 15. pp. 173 - 189.

Ayalew, S. M. (2007). Flood forecasting and early warning system (FFEWS) an alternative technology for flood management system and damage reduction in Ethiopia: A concept note. *Catchment and Lake Research, LARS*, 2007.

Baethgen, W. E., Meinke, H. & Gimenez, A. (2004). Adaptation of agricultural production systems to climate variability and climate change: Lessons learned and proposed research approach. In: *Insights and tools for adaptation: Learning from climate variability*, NOAA-OGP, Washington, D.C

Bahinipati, C. S. (2009). Vulnerability to climate change: how to mainstream adaptation. *Paper Presented at 7th International Conference on the Human Dimensions of Global Environmental Change,* World Conference Center, UN Campus, Bonn. http://www.openmeeting2009.org/pdf_files/Pdf%20papers/Chandra%20shekhar%20bahinipati.pdf.

Barnett, T., Blas, E., & Alan, W., (Eds.). (1996). *AIDS Briefs. Integrating HIV/AIDS into sectoral planning.* World Health Organization, Global Programme on AIDS, Support for Analysis and Research in Africa (SARA) Project, Health and Human Resources Analysis for Africa (HHRAA) Project. U.S. Agency for International Development, Africa Bureau, Office of Sustainable Development.

Basu, J. P. (2011). Adaptation to climate change, vulnerability and micro-insurance business: A study on forest dependent communities in drought prone areas of West Bengal, India. *Working Paper* No. 2011/14, Maastricht School of Management.

Bekele, Selome & Hailemariam, Assefa (2010). Population dynamics and environment in Ethiopia: An overview. In S. Edwards (ed.) *Ethiopian Environment Review,* No. 1. Forum for Environment.

Berkhout, F., Hertin, J. & Gann, D. M. (2004). Learning to adapt: Organizational adaptation to climate change impacts. Tyndall Centre for Climate Change Research, *Working Paper* 47.

Block, L. G. & Keller, P. A. (1998). Beyond protection motivation: An integrative theory of health appeals, *Journal of Applied Social Psychology,* Vol. 28, No. 17, pp. 1584-1608.

Bočkarjova, M., van der Veen, A. & Geurts, P.A.T.M. (2009). Flood disaster in the Netherlands: a trade-off between paying for protection and undertaking action? *Working Papers Series,* Paper 4, International Institute for Geo-Information Science and Earth Observation, University of Twente.

Bodansky, D. (2001). The history of the global climate change regime. In U. Luterbacher, & D. F. Sprinz (Eds.). *International relations and global climate change,* The MIT Press, Massachusetts.

Böhringer, C. & Finus, M. (2005). The Kyoto Protocol: Success or failure? In D. Helm (Ed.) *Climate change Policy.* Oxford University Press Inc., New York.

Bolin, B. (2007). *A History of the science and politics of climate change. The role of the Intergovernmental Panel on Climate Change.* Cambridge University Press, Cambridge.

Brooks, N. & Adger, W. N. (2005). Assessing and enhancing adaptive capacity. In B. Lim and E. Spanger-Siegfried, I. Burton, E. Malone, S. Huq (eds.) *Adaptation Policy Frameworks for Climate Change: Developing Strategies, Policies and Measures*, pp. 165-181. UNDP-GEF. Cambridge University Press.

Brooks, N. (2003). Vulnerability, risk and adaptation: A Conceptual Framework. *Working Paper* 38, Tyndall Centre for Climate Change Research, University of East Anglia, Norwich.

Bryman, A (2006). Integrating quantitative and qualitative research: how is it done? *Qualitative Research*, Vol. 6(1) 97–113.

Burroughs, W. J. (2007). *Climate change: A multidisciplinary approach* (2nd ed.). Cambridge University Press. Cambridge.

Burton, I., Diringer, E. & Smith, J. (2007). Adaptation to Climate Change: International Policy Options. *Paper prepared for the Pew Center on Global Climate Change.*

Cabas, J., Weersink, A. & Olale, E. (2010). Crop yield response to economic, site and climatic variables. *Climatic Change,* 101:599-616.

Camberlin P. (2009). Nile Basin climates. In H. J. Dumont, (Ed.), *The Nile: Origin, environments, limnology and human use,* Monographiae Biologicae, Vol. 89, Springer, 307-333.

Chanyalew, D., Adenew, B. & Mellor, J. (2010). *Ethiopia's Agricultural Sector Policy and Investment Framework (PIF) 2010-2020.* Ministry of Agriculture and Rural Development, Federal Democratic Republic of Ethiopia, Addis Ababa.

Charles, B. (2004). Farm level adoption decisions of soil and water management technologies in semi-arid Eastern Kenya. *Paper presented at the 48th Annual Conference of the Australian Agricultural and Resource Economics Society,* February 11-13, 2004, Melbourne, Australia.

Chen, C., Mccarl, B. A. & Schimmelpfennig, D. E. (2004). Yield variability as influenced by climate: A statistical investigation. *Climatic Change* 66: 239–261.

CLI (2009). *Setting the record straight: Responses to common challenges to climate science.* Climate Leadership Initiative, Institute for a Sustainable Environment, University of Oregon.

Cobb, A. (2011). *Incorporating indigenous knowledge systems into climate change discourse. Colorado Conference on Earth System Governance: Crossing Boundaries and Building Bridges.* Department of Sociology, Colorado State University. http://cc2011.earthsystemgovernance.org/pdf/2011Colora_0130.pdf.

Collins, F.C., & Bolstad, P.V. (1996). A comparison of spatial interpolation techniques in temperature estimation. In: *Proceedings of the Third International Conference/Workshop on Integrating GIS and Environmental Modeling*, Santa Fe, New Mexico, January 21-25, 1996. Santa Barbara, California: National Center for Geographic Information Analysis (NCGIA). CD-ROM.

Coppock, D. L. (ed.) (1994). *The Borana Plateau of Southern Ethiopia: Synthesis of pastoral research, development and change, 1980-91*. ILCA (International Livestock Centre for Africa), Addis Ababa.

CSA (2008). *Summary and statistical report of the 2007 population and housing census results: Population size by age and sex.* Federal Democratic Republic of Ethiopia Population Census Commission, Addis Ababa.

CSA (2011). *Agricultural sample survey 2010 / 2011 (2003 E.C.): Report on area and production of major crops (private peasant holdings, meher season), Vol. I.* Central Statistical Agency, the Federal Democratic Republic of Ethiopia, Addis Ababa.

CSA (2012). *Agricultural sample survey 2011 / 2012 (2004 E.C.): Report on land utilization (private peasant holdings, meher season), Vol. IV.* Central Statistical Agency, the Federal Democratic Republic of Ethiopia, Addis Ababa.

CSA & ORC Macro (2006). *Ethiopia Demographic and Health Survey 2005.* Addis Ababa, Ethiopia and Calverton, Maryland, USA: Central Statistical Agency and ORC Macro.

CSA, EDRI & IFPRI (2006). *Atlas of the Ethiopian rural economy*. Central Statistical Agency (Addis Ababa), Ethiopian Development Research Institute (Addis Ababa), and International Food Policy Research Institute (Washington, DC).

Cutter, S. L., Emrich, C. T., Webb, J. J. & Morath, D. (2009). *Social vulnerability to climate variability hazards: A review of the literature.* Final Report to Oxfam America. Hazards and Vulnerability Research Institute, University of South Carolina, Columbia.

Darwin, R. (1999). A farmer's view of the Ricardian approach to measuring agricultural effects of climatic change. *Climatic Change* 41: 371-411.

De Silva, R. P., Dayawansa, N.D.K. & Ratnasiri, M. D. (2007). A comparison of methods used in estimating missing rainfall data. *The Journal of Agricultural Sciences,* Vol.3, no.2.

Depledge, J. (2004). From negotiation to implementation: The UN Framework Convention on Climate Change and its Kyoto Protocol. *In* A. D. Owen & N. Hanley, *The Economics of Climate Change.* Routledge.

Dercon, S. (2004). Growth and shocks: evidence from rural Ethiopia. *Journal of Development Economics 74:309-329.*

Deressa, T. T. (2007) Measuring the economic impact of climate change on Ethiopian agriculture: Ricardian approach. *Policy Research Working Paper 4342,* The World Bank Development Research Group.

Deressa, T. T. (2010). *Assessment of the vulnerability of Ethiopian agriculture to climate change and farmers' adaptation strategies.* PhD thesis, University of Pretoria.

Deressa, T., Hassan, R. M., Alemu, T., Yesuf, M. & Ringler, C. (2008). Analyzing the Determinants of Farmers' Choice of Adaptation Methods and Perceptions of Climate Change in the Nile Basin of Ethiopia. *IFPRI Discussion Paper 00798.*

Deressa, T., Hassan, M. R., Ringler, C., Alemu, T. & Yusuf, M. (2009). Determinants of farmers' choice of adaptation methods to climate change in the Nile Basin of Ethiopia. *Global Environmental Change,* Vol. 19, No. 2, pp. 248-255.

Deschenes, O. & Greenstone, M. (2006).The Economic impacts of climate change: Evidence from agricultural profits and random fluctuations of weather. *MIT Joint Program on the Science and Policy of Global Change, Report No. 131.*

Di Falco, S., Chavas, J-P. & Smale, M. (2006). Farmer Management of Production Risk on Degraded Lands: The Role of Wheat Genetic Diversity in Tigray Region, Ethiopia. International Food Policy Research Institute. *EPT Discussion Paper 153.*

Dinar, A. & Beach, H. (1998). Measuring the Impact of and the Adaptation to Climate Change in Agriculture and Other Sectors - Literature Review. In Dinar, A., Mendelsohn, R., Evenson, R., Parikh, J., Sagnhi, A., Kumar, K., McKinsey, J. & Lonergan, S. (1998) (Eds) *Measuring the Impact of Climate Change on Indian Agriculture.* World Bank Technical Paper No. 402, The World Bank.

Dougherty, C. (2001). *Introduction to econometrics.* (3rd ed.). Oxford University Press.

Dwyer, A., Zoppou, C., Nielsen, O., Day, S. & Roberts, S. (2004). Quantifying social vulnerability: A methodology for identifying those at risk to natural hazards. *Geoscience Australia Record* 2004/14.

Dynes, R. R. (2002). The importance of social capital in disaster response. *Preliminary paper, No. 327,* Disaster Research Center, Department of Sociology/Criminal Justice, University of Delaware, Newark.

Eckaus, R. S. & Tso, K. K. (1999). A study of the effects of natural fertility, weather and productive inputs in Chinese agriculture. *MIT Joint Program on the Science and Policy of Global Change. Report No. 50,* Massachusetts Institute of Technology, Cambridge.

Ellis, F. (1989). *Peasant economics: Farm households and agrarian development.* (2nd ed.). Cambridge: Cambridge University Press.

EM-DAT, EM-DAT (http://www.emdat.be/result-country-profile). *The OFDA/CRED international disaster database,* www.em-dat.net - Université Catholique de Louvain - Brussels - Belgium.

EPA /Environmental Protection Authority/ (1998). *National Action Programme to Combat Desertification.* Federal Democratic Republic of Ethiopia, Addis Ababa.

Eriksen, S. (2005). *The role of indigenous plants in household adaptation to climate change: the Kenyan experience.* In Pak Sum Low (ed.). *Climate change and Africa.* Cambridge university press, Cambridge.

ERM /Environmental Resource Management/ (2003). *Environmental assessment management framework for the pastoral community development project.* Federal Democratic Republic of Ethiopia, Addis Ababa.

ESPERE Climate Encyclopaedia, English full version 2004-2006. http://espere.mpch-mainz.mpg.de/documents/pdf/Encyclopaediamaster.pdf.

Exenberger, A. & Pondorfer, A. (2011). Rain, temperature and agricultural production: The impact of climate change in Sub-Sahara Africa, 1961-2009. *Working Papers in Economics and Statistics 2011-26,* University of Innsbruck.

FAO (2005). *Irrigation in Africa in figures - AQUASTAT Survey 2005.* Food and Agriculture Organisation of the UN, Rome.

FAO (2008). *Climate change adaptation and mitigation in the food and agriculture sector.* Technical Background Document from the Expert Consultation Held On 5 To 7 March, Rome.

Fezzi, C. & Bateman, I. (2012). Non-linear effects and aggregation bias in Ricardian models of climate change. *CSERGE Working Paper 2012-02*, School of Environmental Sciences, University of East Anglia.

Field, A., (2005).*Discovering statistics using SPSS*. (2nd ed.). Sage Publications.

Fleming, J. R. (1998). *Historical perspectives on climate change*. Oxford University Press, Oxford.

Floyd, D. L., Prentice-Dunn, S. & Rogers, R. W. (2000). A meta-analysis of research on protection motivation theory. *Journal of Applied Social Psychology*, Vol. 30, No.2, pp 407 – 429.

Füssel, H.M. (2007). Adaptation planning for climate change: concepts, assessment approaches, and key lessons. *Sustain Sci*. 2:265–275.

Füssel, H-M. & Klein. R. J. T. (2002). *Assessing vulnerability and adaptation to climate change:* An evolution of conceptual thinking paper presented at the UNDP expert group meeting on "integrating disaster reduction and adaptation to climate change". Havana, Cuba, 17–19 June 2002. http://www.pik-potsdam.de/~fuessel/download/undp02_final.pdf.

Füssel, M-H. & Klein, R.J.T. (2006). Climate change vulnerability assessment: An evolutionary of conceptual thinking. *Climate Change* 75:301-329.

Gagnon, C. A. & Berteaux, D. (2009). Integrating traditional ecological knowledge and ecological science: a question of scale. *Ecology and Society* 14(2): 19. [online] URL: http://www.ecologyandsociety.org/vol14/iss2/art19/.

García-Flecha, M. & Viladrich-Grau, M. (2005). The economic relevance of climate variables in agriculture: The Case of Spain. *Economía Agraria y Recursos Naturales*, Vol. 9, 1, pp. 149-180.

Garnaut, R. (2011).*The Garnaut Review 2011 Australia in the global response to climate change*. Cambridge university press, Cambridge.

Gbetibou, G. A. (2009). Understanding farmers' perception and adaptation to climate change and variability. The Case of Limpopo Basin, South Africa. Environment and Production Technology Division, *IFPRI Discussion Paper 00849*.

Gbetibouo, G.A. & Hassan, R.M. (2005). Measuring the economic impact of climate change on major South African field crops: a Ricardian approach. *Global and Planetary Change* 47: 143-152.

Gibson, J. R. & Loh, C. (2011). *Achieving cost-effective mitigation and adaptation*. Civic Exchange, http://unfccc.int/resource/docs/2012/smsn/ngo/175.pdf.

Giddens, A. (2008).*The politics of climate change: National responses to the challenge of global warming*. Policy network paper, London.

Gonfa, L. (1996). Climate classification of Ethiopia. *Meteorological Research Report Series. Vol. 1, No. 3*, Addis Ababa.

Green, D. & Raygorodetsky, G. (2010). Indigenous knowledge of a changing climate. *Climatic Change*, 100:239-242.

Greene, J. C., Caracelli, V. J., & Graham, W. D. (1989). Toward a conceptual framework for mixed-method evaluation designs. *Educational Evaluation and Policy Analysis, Vol.* 11, No. 3: pp. 255-274.

Griliches, Z. (1963). "Estimates of the aggregate agricultural production function from cross-sectional data." *Journal of Farm Economics*, 45:419–28.

Grothmann, T., and Patt, A. (2005). Adaptive capacity and human cognition: the process of individual adaptation to climate change. *Global Environmental Change* 15 (3): 199–213.

Gryseels, G. & Anderson, F. M., (1983). *Research on farm and livestock productivity in the central Ethiopian highlands: Initial results, 1977-1980.* ILCA Research Report, No. 4, International Livestock Centre for Africa.

Gujarati, D.N., (2004). *Basic econometrics.* (4th Ed.). The McGraw-Hill Companies, Avenue of the Americas, New York.

Halsnæs K. & Verhagen, J., (2007). Development based climate change adaptation and mitigation: conceptual issues and lessons learned in studies in developing countries. *Mitigation Adaptation Strategies for Global Change*, Vol. 12, pp. 665-684.

Harmeling, S. & Christoph, B. (2008). *Adaptation to Climate Change - Where do we go from Bali? An Analysis of the Cop13 and the Key Issues on the Road to a New Climate Change Treaty.* Germanwatch e.V. publishers, on line at: http://www.germanwatch.org/klima/adapt08e.htm.

Hartkamp, A.D., De Beurs, K., Stein, A. & White, J.W. (1999). Interpolation Techniques for Climate Variables. *NRG-GIS Series 99-01*. Mexico, D.F.: CIMMYT.

Hassen, R. and Nhemachena, C. (2008). Determinants of African farmers'' strategies for adapting to climate change: Multinomial choice analysis. *AfJARE* Vol. 2, No 1.

Heidhues, F. & Brüntrup, M. (2003). Subsistence agriculture in development: Its role in processes of structural change. *In S. Abele & K. Frohberg, subsistence agriculture in Central and Eastern Europe: How to break the vicious circle? Volume 22,* Institut Für Agrarentwicklung in Mittel-Und Osteuropa (Iamo), Halle, Germany).

Helm, D. (2005). Climate-change policy: A Survey. In D. Helm (ed.), *Climate-change Policy*. Oxford University Press Inc., New York.

Hoffman, U. (2011). Assuring Food Security in Developing Countries under the Challenges of Climate Change: Key Trade and Development Issues of a Fundamental Transformation of Agriculture. *UNCTAD Discussion Paper, No. 201.*

Hosmer, D.W. & Lemeshow, S., (2000). *Applied logistic regression.* (2nd ed.). A Wiley-Interscience Publication, John Wiley & Sons, Inc.

Hulme, M., Doherty, R., Ngara, T., New, M. & Lister, D. (2001). African climate change: 1900-2100. *Climate Research,* 17: 145-168.

Huq, S., Rahman, K., Atiq ., Sokona, Y. & Reid, H. (2003). *Mainstreaming adaptation to climate change in least developed countries (LDCs).* International Institute for Environment and Development, London. http://www.iied.org/pubs/pdfs/10004IIED.pdf.

Hurni, H. (1998). *Agroecological belts of Ethiopia: Explanatory notes on three maps at a scale of 1:1,000,000.* Soil Conservation Research Programme Ethiopia, Research Report, Centre for Development and Environment (University of Bern, Switzerland) and The Ministry of Agriculture (Addis Ababa, Ethiopia).

IGAD-ICPAC (2007). *Climate change and human development in Africa: Assessing the risks and vulnerability of climate change in Kenya, Malawi and Ethiopia.* UNDP, Draft Report.

IPCC (2001). *Climate Change 2001: Impacts, Adaptation, and Vulnerability.* Contribution of Working Group II to the Third Assessment Report of the Intergovernmental Panel on Climate Change, Cambridge University Press, Cambridge.

IPCC (2007a). *Climate change 2007: Synthesis Report.* Contribution of Working Groups I, II and III to the Fourth Assessment Report of the International Panel on Climate Change [Core Writing Team, Pachauri, R.K. and Reisinger, A. (eds.)]. IPCC, Geneva, Switzerland, 104 pp.

IPCC (2007b). *Climate change 2007: Impacts, adaptation and vulnerability. Contribution of Working Group II to the Fourth Assessment Report of the Intergovernmental Panel on Climate Change*, M.L. Parry, O.F. Canziani, J.P. Palutikof, P.J. van der Linden and C.E. Hanson, (eds.), Cambridge University press, Cambridge, UK, 976 pp.

IPCC (2007c). *Climate Change 2007: The Physical Science Basis.* Contribution of Working Group I to the Fourth Assessment Report of the Intergovernmental Panel on Climate Change [Solomon, S., D. Qin, M. Manning, Z. Chen, M. Marquis, K.B. Averyt, M. Tignor and H.L. Miller (eds.)]. Cambridge University Press, Cambridge, United Kingdom and New York, NY, USA, 996 pp.

Irwin, W. & Williams, B. (2010). An ethical defense of global-warming skepticism. *Reason Papers 32* (Fall 2010): 7-27.

Isik, M. & Devadoss, S. (2006). An analysis of the impact of climate change on crop yields and yield variability. *Applied Economics*, 38, 835–844.

Islam, N. & Salim, R. A. (2009). *Can R&D investment offset the negative impact of climate change on agricultural productivity?* Department of Agriculture and Food, Western Australian Agriculture Authority.

Janssen, M. & Elinor O. (2006). Resilience, vulnerability, and adaptation: A Cross-cutting theme of the International Human Dimensions Programme on Global Environmental Change. *Global Environmental Change* 16:237–239.

Jasanoff, S. (2010). A New climate for society. *Theory, Culture and Society*, Vol. 27(2-3): 233-253.

Jayne, T.S., Yamano, T., Weber, M.T., Tschirley, D., Benfica, R., Chapoto, A., & Zulu, B. (2003) Smallholder income and land distribution in Africa: Implications for poverty reduction strategies. *Food Policy*, Vol. 28, No. 3, pp. 253-275.

Jenkin, C. M. (2006). Risk perception and terrorism: Applying the psychometric paradigm. *Homeland Security Affairs*, Vol. II, No. 2.

Jones, L. (2010). *Overcoming social barriers to adaptation. Background note.* Overseas Development Institute. http://www.odi.org.uk/sites/odi.org.uk/files/odi-assets/publications-opinion-files/6048.pdf.

Jones, R.N., Dettmann, P., Park, G., Rogers, M. & White, T. (2007). The relationship between adaptation and mitigation in managing climate change risks: A Regional response from North Central Victoria, Australia. *Mitigation Adaptation Strategies for Global Change* Vol. 12, Issue 5, pp 685-712.

Just, R.E., & Pope, R.D. (1979). Production Function Estimation and Related Risk Considerations. *American Journal of Agricultural Economics*, 61(2):276-284.

Kabat P., Schulze R.E., Hellmuth M.E. & Veraart J.A. (eds.) (2002). Coping with impacts of climate variability and climate change in water management: A scoping paper. *DWC-Report no. DWCSSO-01(2002)*, International Secretariat of the Dialogue on Water and Climate, Wageningen.

Kabubo-Mariara, J. & Karanja, F. K. (2006). *The economic impact of climate change on Kenyan crop agriculture: A Ricardian approach*. Paper prepared for the Third World Congress of Environmental and Resource Economics: Kyoto Japan, July 3-7, 2006.

Kandji, T. S., Verchot, L., & Mackensen, J. (2006). *Climate Change Climate and Variability in Southern Africa: Impacts and Adaptation in the Agricultural Sector*. UNEP & ICRAF.

Kandlikar, M. & Risbey, J. (2000). Agricultural impacts to climate change: If adaptation is the answer, what is the question? *Climate Change* 45: pp 529 - 539.

Kasperson, R. E., Renn, O., Slovic, P., Brown, H. S., Emel, J., Goble, R., Kasperson, J. X. & Ratick, S. (1988). The social amplification of risk: A conceptual framework. *Risk Analysis*. Vol. 8, No. 2. pp. 177 - 187.

Kates, R.W. (1997). Review of climate change: 1995: Impacts, adaptations, and mitigation. In R.T. Watson, M.C. Zinyowera & R.H. Moss (eds.), *Environment*, Vol. 39 No. 9, PP. 29 - 33.

Kelman, I. & West, J. J. (2009). Climate change and Small Island Developing States: A critical review. *Ecological and Environmental Anthropology*, Vol. 5, No. 1.

Kirkland, E. (2012). *Indigenous knowledge and climate change adaptation in the Peruvian Andes*. Brown's Climate and Development Lab.

Klein, R. J. T and MacIver, D. (1999). Adaptation to climate variability and change: Methodological issues. *Mitigation and Adaptation Strategies for Global Change 4: 189-198.*

Klein, R. J. T. & Person, Åsa (2008). Financing Adaptation to Climate Change: Issues and Priorities. European Climate Platform (ECP). An Initiative of Mistra's Climate Policy Research Programme (Clipore) and the Centre for European Policy Studies (CEPS), *ECP Report No. 8*.

Klein, R.J.T. (2001). *Adaptation to climate change in German Official Development Assistance: An inventory of activities and opportunities, with a special focus on Africa.* Deutsche Gesellschaft für Technische Zusammenarbeit, Eschborn, Germany, 42 pp.

Kloos, H. & Adugna, A. (1989). The Ethiopian population: Growth and distribution. *The Geographical Journal*, Vol. 155, No. 1, pp. 33-51.

Koenker, R. & Hallock, K. F., (2001). Quantile regression. *Journal of Economic Perspectives*, Vol. 15, No. 4, pp. 143–156.

Koenker, R.W. & Bassett Jr., G.S., (1978). Regression quantiles. *Econometrica*, Vol. 46, No. 1, pp. 33-50.

Kopeva, D. & Noev, N. (2003). Subsistence Farming in Bulgaria: Between Tradition and Market Requirements. In Steffen Abele and Klaus Frohberg (Eds.), *Subsistence Agriculture in Central and Eastern Europe: How to Break the Vicious Circle?* Institute of Agricultural Development in Central and Eastern Europe, Volume 22.

Koundouri, P. & Nauges, C. (2005). On Production Function Estimation with Selectivity and Risk Considerations. *Journal of Agricultura1 and Resource Economics*, 30(3):597-608.

Kurukulasuriya, P. & Mendelsohn, R., (2007). A Ricardian Analysis of the Impact of Climate Change on African Cropland. *World Bank Policy Research Working Paper No. 4305.*

Kurukulasuriya, P. & Rosenthal, S. (2003). Climate change and agriculture: A review of impacts and adaptations. *Climate Change Series*, Paper No. 91. The World Bank Environment Department. The World Bank, Washington D.C.

Leggett, J. A. & Lattanzio, R. K. (2009). *Status of the Copenhagen climate change negotiations.* CRS Report for Congress, Congressional Research Service, http://fpc.state.gov/documents/organization/135974.pdf.

Leiserowitz, A. (2007). Public perception, opinion and understanding of climate change - current patterns, trends and imitations. *Human Development Report 2007/2008.* UNDP.

Levina, E. (2007). *Adaptation to climate change: International agreements for local needs.* Organization for economic Co-Operation and Development, OECD/IEA, Paris.

Levina, E. and Tirpak, D. (2006). Key adaptation concept and terms. *Draft Paper/Agenda Document 1,* OECD/IEA Project for the Annex I Expert Group on the UNFCC. http://www.eird.org/cd/on-better-terms/docs/Organisation-of-Economic-Co-operation-and-Development.pdf.

Ligeon, C., Jolly, C., Bencheva, N., Delikostadinov, S. & Puppala, N. (2008). Production risks in Bulgarian peanut production. *Agricultural Economics Review,* Vol. 9, No 1.

Lomborg, B. (2006). *The skeptical environmentalist: Measuring the real state of the world.* 7th Printing, Cambridge University Press, Cambridge.

Lotze-Campen, H. & Schellnhuber, H. J. (2009). Climate impacts and adaptation options in agriculture: what we know and what we don't know. - Journal für Verbraucherschutz und Lebensmittelsicherheit - *Journal for Consumer Protection and Food Safety,* Vol. 4, Issue 2, pp. 145-150.

Ludwig, F., van Scheltinga, C. T., Verhagen, J., Kruijt, B., van Ierland, E., Dellink, R., de Bruin, K., de Bruin, K. & Kabat, P. (2007). *Climate change impacts on Developing Countries - EU Accountability.* Policy Department, Economic and Scientific Policy, The European Parliament.

Luo, W., Taylor, M. C. & Parker, S. R. (2008). A comparison of spatial interpolation methods to estimate continuous wind speed surfaces using irregularly distributed data from England and Wales. *International Journal of Climatology.* 28: 947–959.

Lupton, D. (2006). *Risk.* London: Routledge.

Macchi, M., Oviedo, G., Gotheil, S., Cross, K., Boedhihartono, A., Wolfangel, C. & Howell, M. (2008). *Indigenous and traditional peoples and climate change. Issues Paper.* IUCN.

Maddala, G.S. (1992). *Introduction to econometrics.* (2nd ed.). Macmillan Publishing Company, New York.

Maddala, G.S., (1983). *Limited-dependent and qualitative variables in econometrics.* Cambridge University Press, Cambridge.

Maddison, D. (1998). The amenity value of climate: The household production function approach. *Resource and Energy Economics,* Vol. 25 Issue 2, pp.155-175.

Maddison, D. (2006). The perception of and adaptation to climate change in Africa. *Policy Research Working Paper 4308,* Development Research Group, The World bank, Washington DC.

Maddison, D., Manley, M. & Kurukulasuriya, P. (2007).The Impact of climate change on African agriculture: A Ricardian approach. *Policy Research Working Paper 4306*, The World Bank, Washington DC.

Majumder, M., Pramanik, S., Barman, Rabindra N., Roy, P. & Mazumdar, A. (2010). Impact of climate change on the availability of virtual water estimated with the help of distributed neurogenetic models. In B. K. Jana and M. Majumder (Eds.), *Impact of climate change on natural resource management*. Springer.

Mano, R. & Nhemachena, C. (2006). Assessment of the economic impacts of climate change on agriculture in Zimbabwe: a Ricardian approach. *CEEPA Discussion Paper No. 11*, Centre for Environmental Economics and Policy in Africa, University of Pretoria.

Martin, I.M., Bender, H. W. & Raish, C. (2008). Making the decision to mitigate risk. In W. E. Martin, C. Raish, and B. Kent (Eds.), *Wildfire risk : human perceptions and management implications*. Washington, DC: Resources for the Future, p. 117-141.

McCarl, B. A., Adams, R. M. & Hurd, B. H. (2001). *Global climate change and its impact on agriculture.* http://agecon2.tamu.edu/people/faculty/mccarl-bruce/papers/879.pdf.

McGray, H., Hammill, A. & Bradley, R. (2007). *Weathering the Storm. Options for Framing Adaptation and Development*. World Resources Institute, Washington, DC.

McGuigan, C., Reynolds, R., & Wiedmer, D. (2002). *Poverty and Climate Change: Assessing Impacts in Developing Countries and the Initiatives of the International Community.* London School of Economics Consultancy Project for the Overseas Development Institute.

Mearns, L. O., Rosenzweig, C. & Goldberg, R. (1997). Mean and variance change in climate scenarios: Methods, agricultural applications, and measures of uncertainty. *Climatic Change* 35: 367-396.

Meinzen-Dick, R., Quisumbing, A., Behrman, J., Biermayr-Jenzano, P., Wilde, V., Noordeloos, M., Ragasa, C., & Beintema, N. (2010). Engendering Agricultural Research. *IFPRI Discussion Paper 00973.*

Menard, S. (2001). *Applied Logistic Regression Analysis*. Sage University Papers Series on Quantitative Applications in the Social Sciences, 07-106, Thousand Oaks, CA, Sage.

Mendelsoh, R. & Reinsborough, M. (2007). A Ricardian analysis of US and Canadian farmland. *Climatic Change*, 81:9-17.

Mendelsohn, R. & Dinar, A. (1999). Climate Change, Agriculture, and Developing Countries: Does Adaptation Matter? *The World Bank Research Observer*. Vol. 14, no. 2, pp. 277-93.

Mendelsohn, R. & Dinar, A. (2003). Climate, water, and agriculture. *Land Economics* 79(3): 328–341.

Mendelsohn, R., Nordhaus, W. D., & Shaw, D. (1994). The Impact of Global Warming on Agriculture: A Ricardian Analysis. *The American Economic Review*, Vol. 84, No. 4, pp. 753-771.

Mengistu, A. (2003). *Country pasture/forge resource profile: Ethiopia.* http://www.fao.org/ag/AGP/AGPC/doc/counprof/ethiopia/Ethiopia.htm#figure5.

Menzel, S. & Scarpa, R. (2005). Protection motivation theory and contingent valuation: Perceived realism, threat and WTP estimates for biodiversity protection. *FEEM Working Paper No. 26.05.*

Mileti, D. S. (1993). Communicating public earthquake risk information. In J. Nemec, J.M. Nigg & F.Siccardi (Eds.), *Prediction and Perception of Natural Hazards*, pp. 143 – 152. Kluwer Academic Publishers.

MOFED /Ministry of Finance and Economic Development/ (2003). *Rural development policy and strategies.* Economic Policy and Planning Department, Ministry of Finance and Economic Development, Government of the Federal Democratic Republic of Ethiopia, Addis Ababa.

Moffatt, I. (2004). Global warming: Scientific modeling and its relationship to the economic dimensions of policy. In A. D. Owen and N. Hanley, *The economics of climate change.* Routledge.

MOH (2008). *Ethiopian national malaria indicator survey 2007.* Addis Ababa, Ethiopia.

Molua, E. & Lambi, C. (2006). The economic impact of climate change on agriculture in Cameroon. *CEEPA Discussion Paper No. 17*, Centre for Environmental Economics and Policy in Africa, University of Pretoria.

Mulilis, J-P & Lippa, R. (1990). Behavioral change in earthquake preparedness due to negative threat appeals: A test of Protection Motivation Theory. *Journal of Applied Social Psychology*, Vol. 20, No. 8, pp. 619-638.

Nabikolo, D., Bashaasha, B., & Mangheni, M.N. (2012). Determinants of climate change adaptation among male and female headed farm households in Eastern Uganda. *African Crop Science Journal*, Vol. 20, Issue Supplement s2, pp. 203 - 212.

Nakashima, D.J., Galloway McLean, K., Thulstrup, H.D., Ramos Castillo, A. & Rubis, J.T. (2012). *Weathering Uncertainty: Traditional Knowledge for Climate Change Assessment and Adaptation*. Paris, UNESCO, and Darwin, UNU, 120 pp.

NAS/National Academy of Sciences (2010). *Advancing the science of climate change*. National Academies Press, Washington, DC.

Neu, U. (2009). *Climate skeptic arguments and their scientific background*. Swiss Reinsurance Company Ltd. Zurich. http://proclimweb.scnat.ch/portal/ressources/1183.pdf.

Nhemachena, C. & Hassan, R. (2007). Micro-level analysis of farmers' adaptation to climate change in Southern Africa. *IFPRI Discussion Paper 00714*.

NMSA (1996). Climatic and Agroclimatic Resources of Ethiopia. *Meteorological Research Report Series*, Vol. 1, No. 1, Addis Ababa.

NMSA (2001). *Initial National Communication of Ethiopia to the United Nations Framework Convention on Climate Change (UNFCCC)*. Federal Democratic Republic of Ethiopia, Ministry of Water Resources and National Meteorological Services Agency, Addis Ababa.

Norman, P., Boer, H. & Seydel, E. R. (2005). *Protection Motivation Theory*. In M. Conner & P. Norman (Eds.), *Predicting Health Behavior: Research and Practice with Social Cognition Models*. (2nd ed.). pp. 81-126. Open University Press, Maidenhead.

Nyong, A., Adesina, F. & Elasha, B. O. (2007). The value of indigenous knowledge in climate change mitigation and adaptation strategies in the African Sahel. *Mitigation Adaptation Strategies for Global Change*, Vol. 12:787–797.

O'Brien, K.L. & Leichenko, R.M. (2000). Double exposure: assessing the impacts of climate change within the context of economic globalization. *Global Environmental Change*, vol. 10, pp. 221-232.

Oboh, V.U. & Kushwaha, S. (2009). Socio-economic determinants of farmers' loan size in Benue State, Nigeria. *Journal of Applied Sciences Research*, 5(4), 354-358.

Onemolease, E. A. & Alakpa, S. O. (2009). Determinants of adoption decisions of rural youths in the Niger Delta Region of Nigeria. *Journal of Social Sciences*, Vol. 20, No. 1.

Ouedraogo, M., Some, L. & Dembele, Y. (2006). *Economic Impact Assessment of Climate Change on Agriculture in Burkina Faso: A Ricardian Approach.* Centre for Environmental Economics and Policy in Africa (CEEPA).

Oury, B. (1965). Allowing for Weather in Crop Production Model Building . *Journal of Farm Economics,* Vol. 47, No. 2.

Padel, S. (2001). Conversion to organic farming: A typical example of the diffusion of an innovation? *Sociologia Ruralis,* Vol. 41 No. 1.

Pampel F.C., (2000). *Logistic Regression: A Primer.* Sage University Papers Series on Quantitative Applications in the Social Sciences, 07-132, Thousand Oaks, CA, Sage.

Pearson, L. & Langridge, J. (2008). Climate change vulnerability assessment: Review of agricultural productivity, *CSIRO Climate Adaptation Flagship Working paper No.1.*

Pechmann, C., Zhao, G., Goldberg, M. E. & Reibling, E. T. (2003). What to convey in antismoking advertisements for adolescents: The use of Protection Motivation Theory to identify effective message themes. *Journal of Marketing,* 67: 1-18.

Pelling, M. (2011). Climate change and social capital. *Discussion paper 5,* a Report commissioned as part of the UK Government's Foresight Project on the International Dimensions of Climate Change, Government Office for Science. London.

Peng, C. J., Lee, K. L. & Ingersoll, G. M., (2002). An Introduction to Logistic Regression: Analysis and Reporting. *The Journal of Educational Research.* Vol. 96, No. 1.

Petermann, T. (ed.) (2008). *Towards climate change adaptation - Building adaptive capacity in managing African transboundary river basins. Case studies from African practitioners and researchers.* InWEnt - Internationale Weiterbildung und Entwicklung gGmbH Capacity Building International, Germany.

Philander, S. G. (general ed.) (2008). *Encyclopedia of global warming and climate change volumes 1 - 3,* SAGE Publications, Inc.

Pielke, R.A.J. (1998). Rethinking the role of adaptation in climate policy. *Global Environmental Change* 8, 159–170.

Pittock, A. B. (2009). *Climate change: The Science, impacts and solutions.* (2nd ed.). CSIRO publishing, Collingwood.

Plant Genetic Resources Center (1995). *Ethiopia: Country report to the FAO International Technical Conference on plant genetic resources (Leipzig, 1996)*, Addis Ababa, April 1995.

Plapp, T. and U. Werner (2006). Understanding risk perception from natural hazards: examples from Germany. In: D. Amman & Vulliet (Eds.), *RISK 21 - Coping with risks due to natural hazards in the 21st century*, pp. 101-108, Taylor & Francis Group, London.

Preston, B.L. and Stafford-Smith, M. (2009). Framing vulnerability and adaptive capacity assessment: Discussion paper. *CSIRO Climate Adaptation Flagship Working paper No. 2.*

Quiroga, S., Fern´andez-Haddad, Z., & Iglesias, A. (2011). Crop yields response to water pressures in the Ebro basin in Spain: risk and water policy implications. *Hydrology and Earth Systems Science*, 15, 505–518.

Ramsey, J. B. (1969). Tests for specification errors in classical linear least-squares regression analysis. *Journal of the Royal Statistical Society*, Series B (Methodological), Vol. 31, Issue 2, pp. 350–371.

Raygorodetsky, G. (2011). *Why Traditional Knowledge Holds the Key to Climate Change.* United Nations Univeisty, http://unu.edu/publications/articles/why-traditional-knowledge-holds-the-key-to-climate-change.html.

Reilly, J., et al., (2001). *Agriculture: The Potential Consequences of Climate Variability and Change for the United States.* US National Assessment of the Potential Consequences of Climate Variability and Change, US Global Change Research Program. Cambridge University Press, New York, NY, 136 pp.

Rezvanfar, A., Samiee, A. & Faham, E. (2009). Analysis of factors affecting adoption of sustainable soil conservation practices among wheat growers. *World Applied Sciences Journal*, Vol. 6, No. 5, pp. 644-651.

Ribot, J. C. (2009). Vulnerability does not just fall from the sky: Toward multi-scale pro-poor climate policy. In: R. Mearns & A. Norton (Eds.), *Social dimensions of climate change: Equity and vulnerability in a warming world.* The World Bank, Washington, DC.

Richter, B. (2010). *Beyond smoke and mirrors. Climate change and energy in the 21st century.* Cambridge University Press, Cambridge.

Risbey, J., Kandlikar, M., Dowlatabadi, H., & Graetz, D. (1999). Scale, context and decision-making in agricultural adaptation to climate variability and change. *Mitigation and Adaptation Strategies for Global Change* 4 (2), 137–165.

Rohde, R., Curry, J., Groom, D., Jacobsen, R., Muller, R., Perlmutter, A. S., Rosenfeld, A., Wickham, C. & Wurtele, J. (2011). *Berkeley Earth Temperature Averaging Process.* University of California, Berkeley.

Rosegrant, M. W. & Evenson, R. E., (1995). Total factor productivity and sources of long-term growth in Indian agriculture. Environment and Production Technology Division (EPTD), *Discussion Paper No. 7*, International Food Policy Research Institute, Washington, D.C.

Rosenzweig, C. & Hillel, D. (2008). *Climate variability and the global harvest: Impacts of El Niño and other oscillations on Agroecosystems.* Oxford University Press, Inc., Oxford.

Rossi, P.E. (1984). Stochastic specification of cost and production relationships. *Discussion Paper No. 616.* J.L. Kellogg Graduate School of Management, Northwestern University, Evanston, Illinois.

Rowhani, P., Lobell, D. B., Linderman, M. & Ramankutty, N. (2010). Climate variability and crop production in Tanzania. *Agricultural and Forest Meteorology* 151: 449–460.

Saha, A., Havenner, A., & Talpaz H. (1997). Stochastic production function estimation: Small sample properties of ML versus FGLS. *Applied Economics* 29:459-469.

Salick, J. & Byg, A. (2007). *Indigenous peoples and climate change.* Tyndall Centre for Climate Change Research, A Tyndall Centre Publication, Oxford.

Salvatore, D. & Reagle, D. (2002). *Theory and problems of statistics and econometrics.* (2nd ed.). Schaum's Outline Series, The McGraw-Hill Companies.

Sánchez, J. G. & Hernández, O. N. (1995). *Perception of risk by the residents of a flood hazard area in Puerto Rico. Phase 1* . Final Technical Report to U.S Department of the Interior Geological Survey.

Schimmelpfennig, D., Lewandrowski, J., Reilly, J., Tsigas, M. & Parry, I. (1996). Agricultural adaptation to climate change: Issues of long run sustainability. *Agricultural Economic Report,* No. (AER740), 68 pp.

Schipper, E.L.F. (2007). Climate Change Adaptation and Development: Exploring the Linkages. *Tyndall Centre Working Paper No.107.*

Schlickenrieder, J., Quiroga, S., Diz1, A. & Iglesias, A. (2011). Impacts and adaptive capacity as drivers for prioritizing agricultural adaptation to climate change in Europe. *Economía Agraria y Recursos Naturales.* Vol. 11,1. pp.59-82.

Segerson K., & Dixon, B. L. (1998). Climate change and agriculture: the role of farmer adaptation. In Mendelsohn R, Neumann J (Eds.), *The economic impacts of climate change on the U.S. economy*. Cambridge University Press, Cambridge.

Seo, S. N., Mendelsohn, R., Dinar, A., Hassan, R. & Kurukulasuriya, P. (2008). A Ricardian analysis of the distribution of climate change impacts on agriculture across agro-ecological zones in Africa. *Policy Research Working Paper 4599*, the World Bank, Washington DC.

Seo1, S. N. & Mendelsohn, R. (2008a). A structural Ricardian analysis of climate change impacts and adaptations in African agriculture. *Policy Research Working Paper 4603*, The World Bank, Washington DC.

Seo1, S. N. & Mendelsohn, R. (2008b). A Ricardian analysis of the impact of climate change on South American farms. *Chilean Journal of Agricultural Research* 68(1):69-79.

Seo1, S. N., Mendelsohn, R. & Munasinghe, M. (2005). Climate change and agriculture in Sri Lanka: a Ricardian valuation. *Environment and Development Economics* 10:581-596.

Smit, B. & Pilifosova, O. (2001). Adaptation to Climate Change in the Context of Sustainable Development and Equity. In *Climate change 2001: Impacts, adaptation, and vulnerability* - Contribution of Working Group II to the Third Assessment Report of the Intergovernmental Panel on Climate Change. Cambridge University Press, Cambridge.

Smit, B. & Skinner, M. W. (2002). Adaptation options in agriculture to climate change: A typology. *Mitigation and Adaptation Strategies for Global Change* 7: 85 - 114.

Smit, B. & Wandel, J. (2006). Adaptation, adaptive capacity and vulnerability. *Global Environment Change 16: 282 - 292.*

Smit, B., Burton, I., Klein, R.J.T. & Street, R. (1999). The science of adaptation: a framework for assessment. *Mitigation and Adaptation Strategies for Global Change* 4, pp. 199-213.

Sperling, F. & Szekely, F., (2005). *Disaster risk management in a changing climate.* Discussion Paper prepared for the World Conference on Disaster Reduction on behalf of the Vulnerability and Adaptation Resource Group (VARG). Reprint with Addendum on Conference outcomes. Washington, D.C.

SPSS (2007). *SPSS statistics base 17.0.* User's guide. SPSS Inc. Chicago.

Tabachnick, B.G., & Fidell, L.S., (2007). *Using multivariate statistics.* (5th ed.). Pearson Education. Inc.

Tadege, Abebe (ed.) (2007). *Climate change National Adaptation Programme of Action (NAPA) of Ethiopia.* National Meteorological Agency, Ministry of Water Resources, Federal Democratic Republic of Ethiopia, Addis Ababa.

Thapa , S. & Joshi, G. R., (2010). A Ricardian analysis of the climate change impact on Nepalese agriculture. *MPRA Paper No. 29785.*

Tiamiyu S.A., Akintola J.O., & Rahji M.A.Y. (2010). Production efficiency among growers of new rice for Africa in the Savanna Zone of Nigeria. *Agricultura Tropica Et Subtropica.* Vol. 43 (2).

Torvanger, Asbjørn, Twena, Michelle and Romstad, Bård (2004). Climate Change Impacts on Agricultural Productivity in Norway. *CICERO Working Paper 2004:10.*

Tuckett, R. P. (2009). The role of atmospheric gases in global warming. In T. M. Letcher (Ed.), *Climate change: Observed impacts on planet earth.* Elsevier, Amsterdam.

UNCTAD (2010). Agriculture at the crossroads: Guaranteeing food security in a changing global climate. *UNCTAD Policy Briefs,* No. 18.

UNDP (2007). *Human Development Report 2007/2008. Fighting climate change: Human solidarity in a divided world.* United Nations Development Programme. New York.

UNDP (2011). *Human Development Report 2011. Sustainability and equity: A better future for all.* The UNDP Human Development Report Office, New York.

UNFCCC (1992). *United Nations Framework Convention on Climate Change.* United Nations.

UNFCCC (1998). *Kyoto Protocol to the United Nations Framework Convention on Climate Change.* United Nations.

UNFCCC (2007). *Climate Change: Impacts, Vulnerabilities and Adaptation in Developing Countries*. Produced by the Information Services of the UNFCCC secretariat, Bonn.

UNFCCC (2008). *Kyoto Protocol reference manual: On accounting of emissions and assigned amount*. The Information Services of the UNFCCC Secretariat.

USAID (2007). *Adapting to Climate Variability and Change: A Guidance Manual for Development Planning*. Washington, D.C.

Verbeek, M. (2004). *A Guide to modern econometrics.* (2nd ed.). John Wiley & Sons Ltd, The Atrium, Southern Gate, Chichester, West Sussex PO19 8SQ, England.

Wall, E. & Marzall, K. (2006). Adaptive Capacity for Climate Change in Canadian Rural Communities. *Local Environment,* Vol. 11, No. 4, 373-397.

Wallington, T. J., Srinivasan, J., Nielsen, O. J., Highwood, E. J. (2009). Greenhouse gases and global warming. *Environmental and Ecological chemistry, Vol. I.*

Warr, P. (2012). *Research and agricultural productivity in Indonesia.* Paper contributed to Australian Agricultural and Resource Economics Society, 56th conference, Fremantle, February 7 to 10, 2012. Australian National University.

Washington, R., Harrison, M., Conway, D., Black, E., Challinor, A., Grimes, D., Jones, R., Morse, A., Kay, G. & Todd, M. (2006). *African Climate Change: Taking the Shorter Route*. American Meteorological Society.

Wickham, C., Curry, J., Groom, D., Jacobsen, R., Muller, R., Perlmutter, S., Rohde, R., Rosenfeld, A. & Wurtele, J. (2011). *Influence of urban heating on the global temperature land average using rural sites identified from MODIS classifications.* University of California, Berkeley.

Wolde-Georgis, T. (2000). *The case of Ethiopia reducing the impacts of environmental emergencies through early warning and preparedness: The case of the 1997-98 El Niño.* http://archive.unu.edu/env/govern/ElNIno/CountryReports/pdf/ethiopia.pdf.

Woolcock, M. & Narayan, D. (2000). Social capital: Implications for development theory, research, and policy. *The World Bank Research Observer,* Vol. 15, No. 2, pp. 225–49.

Wooldridge, J. M., (2002a). *Econometric analysis of cross-section and panel data.* The MIT Press, Cambridge, Massachusetts.

Wooldridge, J. M., (2002b). *Introductory econometrics: A modern approach.* South-Western College Publishing, 5191 Natorp Boulevard, Mason.

Worku, M. (2007). The missing links: Poverty, population, and the environment in Ethiopia. *Focus on Population, Environment, and Security,* Issue 14, Woodrow Wilson International Center for Scholars and USAID.

World Bank (1998). *Indigenous knowledge for development: A framework for action.* Knowledge and Learning Center, Africa Region. The World Bank, Washington DC.

World Bank (2008). Climate change impacts in drought and flood affected areas: Case studies in India. *Report No. 43946-IN,* Sustainable Development Department Social, Environment and Water Resources Management Unit, India Country Management Unit, South Asia Region.

Zannetti, P. (1998). *Today's debate on global climate change: Searching for the scientific truth.* EnviroComp Institute. www.envirocomp.org.

Ziervogel, G., Cartwright, A., Tas, A., Adejuwon, J., Zermoglio, F., Shale, M. & Smith, B. (2008). *Climate change and adaptation in African agriculture.* Research Report prepared for Rockefeller Foundation by Stockholm Environment Institute.

Appendix

Appendix 1: Chi-Square Test on precipitation perception with all groups

Perception about the trend of precipitation in the last twenty years

	Observed N	Expected N	Residual
Decreasing	207	62.3	144.8
Increasing	34	62.3	-28.3
No change	6	62.3	-56.3
Don't know	2	62.3	-60.3
Total	249		

Test Statistics

	Perception about the trend of precipitation in the last twenty years
Chi-Square	458.550[a]
df	3
Asymp. Sig.	.000

a. 0 cells (.0%) have expected frequencies less than 5. The minimum expected cell frequency is 62.3.

Appendix 2: Chi-Square Test on precipitation perception with two major groups

Frequencies

	Perception about the trend of precipitation in the last twenty years			
	Category	Observed N	Expected N	Residual
1	Decreasing	207	120.5	86.5
2	Increasing	34	120.5	-86.5
Total		241		

Test Statistics

	Perception about the trend of precipitation in the last twenty years
Chi-Square	124.187[a]
df	1
Asymp. Sig.	.000

a. 0 cells (.0%) have expected frequencies less than 5. The minimum expected cell frequency is 120.5.

Appendix 3: Chi-Square Test on temperature perception with all groups

Perception about the trend of temperature in the last twenty years

	Observed N	Expected N	Residual
Decreasing	9	83.0	-74.0
Increasing	239	83.0	156.0
No change	1	83.0	-82.0
Total	249		

Test Statistics

	Perception about the trend of temperature in the last twenty years
Chi-Square	440.193[a]
df	2
Asymp. Sig.	.000

a. 0 cells (.0%) have expected frequencies less than 5. The minimum expected cell frequency is 83.0.

Appendix 4: Chi-Square Test on temperature perception with two major groups

Frequencies

	Perception about the trend of temperature in the last twenty years			
	Category	Observed N	Expected N	Residual
1	Decreasing	9	124.0	-115.0
2	Increasing	239	124.0	115.0
Total		248		

Test Statistics

	Perception about the trend of temperature in the last twenty years
Chi-Square	213.306[a]
df	1
Asymp. Sig.	.000

a. 0 cells (.0%) have expected frequencies less than 5. The minimum expected cell frequency is 124.0.

Appendix 5: Heteroskedasticity and multicollinearity tests results for linear regression in section 5.4 (stage 1 regression)

Heteroskedasticity Test

Breusch-Pagan / Cook-Weisberg test for heteroskedasticity

Ho: Constant variance
Variables: fitted values of CROPROD
chi2(1) = 7.06
Prob > chi2 = 0.0079

Multicollinearity

Variable	VIF	Tolerance (1/VIF)
LAND	1.39	0.717473
FERT	1.34	0.748862
LVSTK	1.30	0.771114
SEED	1.24	0.808619
LABOR	1.22	0.821398
Mean VIF	1.30	

Appendix 6: Heteroskedasticity and multicollinearity test results for linear regression in section 5.4 (stage 2 regression with DVPPT, rainfall deviance from its long year mean)

Heteroskedasticity Test

Breusch-Pagan / Cook-Weisberg test for heteroskedasticity

Ho: Constant variance
Variables: fitted values of rCROPROD
chi2(1) = 9.87
Prob > chi2 = 0.0017

Multicollinearity

Variable	VIF	Tolerance (1/VIF)
AGE	5.43	0.184246
AGREXP	5.29	0.189048
PPT	4.89	0.204628
DVPPT	3.62	0.276002
MNTEMP	3.57	0.280414
DIST	2.39	0.418885
MXTEMP	2.38	0.419764
ACCREDIT	2.02	0.494671
EDUC	1.51	0.663036
EXTN	1.32	0.757129
ADAPT	1.22	0.822845
SEX	1.21	0.826499
NOSO	1.20	0.835780
Mean VIF	2.77	

Appendix 7: Heteroskedasticity and multicollinearity tests results for linear regression in section 5.4 (stage 2 regression with CESS, cessation of rainfall)

Heteroskedasticity Test

Breusch-Pagan / Cook-Weisberg test for heteroskedasticity

Ho: Constant variance
Variables: fitted values of rCROPROD
chi2(1) = 9.87
Prob > chi2 = 0.0017

Multicollinearity

Variable	VIF	Tolerance (1/VIF)
PPT	6.81	0.146897
CESS	6.45	0.155096
AGE	5.43	0.184246
AGREXP	5.29	0.189048
MNTEMP	4.55	0.219719
DIST	2.39	0.418885
MXTEMP	2.33	0.429993
ACCREDIT	2.02	0.494671
EDUC	1.51	0.663036
EXTN	1.32	0.757129
ADAPT	1.22	0.822845
SEX	1.21	0.826499
NOSO	1.20	0.835780
Mean VIF	3.21	

Appendix 8: Linearity and multicollinearity tests and classification table for logistic regression with perceptual variables

Linearity test

Logistic regression		Number of obs	=	217
		LR chi2(2)	=	80.68
		Prob > chi2	=	0.0000
Log likelihood = -102.49641		Pseudo R2	=	0.2824

ADAPT	Coef.	Std. Err.	z	P>|z|	[95% Conf. Interval]
_hat	1.048	0.175	5.990	0.000	0.705 - 1.391
_hatsq	-0.048	0.082	-0.580	0.559	-0.210 - 0.113
_cons	0.055	0.206	0.270	0.791	-0.349 - 0.459

Multicollinearity

Variable	VIF	Tolerance (1/VIF)
SE	3.690	0.271
PC	2.400	0.417
AE	2.190	0.457
PS	1.110	0.903
PV	1.100	0.909
Mean VIF	2.100	

Classification table

Classified	D	~D	Total
+	125	31	156
-	12	49	61
Total	137	80	217

<table>
<tr><td colspan="2">Classified + if predicted Pr(D) >= .5</td></tr>
<tr><td>True D defined as ADAPT != 0</td><td></td></tr>
<tr><td>Sensitivity Pr(+ | D)</td><td>91.24%</td></tr>
<tr><td>Specificity Pr(- | ~D)</td><td>61.25%</td></tr>
<tr><td>Positive predictive value Pr(D | +)</td><td>80.13%</td></tr>
<tr><td>Negative predictive value Pr(~D | -)</td><td>80.33%</td></tr>
<tr><td>False + rate for true ~D Pr(+ | ~D)</td><td>38.75%</td></tr>
<tr><td>False - rate for true D Pr(- | D)</td><td>8.76%</td></tr>
<tr><td>False + rate for classified + Pr(~D | +)</td><td>19.87%</td></tr>
<tr><td>False - rate for classified - Pr(D | -)</td><td>19.67%</td></tr>
<tr><td>Correctly classified</td><td>80.18%</td></tr>
</table>

Appendix 9: Linearity and multicollinearity tests and classification table for logistic regression with socioeconomic and institutional variables

Linearity test

Logistic regression   Number of obs = 217
LR chi2(2) = 57.11
Prob > chi2 = 0.0000
Log likelihood = -114.28165   Pseudo R2 = 0.1999

<table>
<tr><th>ADAPT</th><th>Coef.</th><th>Std. Err.</th><th>z</th><th>P>|z|</th><th>[95% Conf. Interval]</th></tr>
<tr><td>_hat</td><td>1.163</td><td>0.219</td><td>5.310</td><td>0.000</td><td>0.734 - 1.593</td></tr>
<tr><td>_hatsq</td><td>-0.142</td><td>0.118</td><td>-1.200</td><td>0.228</td><td>-0.373 - 0.089</td></tr>
<tr><td>_cons</td><td>0.107</td><td>0.198</td><td>0.540</td><td>0.589</td><td>-0.280 - 0.494</td></tr>
</table>

Multicollinearity

Variable	VIF	Tolerance (1/VIF)
AGE	5.270	0.190
AGREXP	4.690	0.213
ACCREDIT	1.610	0.621
DIST	1.600	0.623
LANDSIZE	1.560	0.640
LVST09	1.460	0.686
EDUC	1.380	0.726
INFOWTH	1.310	0.761
EXTN	1.260	0.794
DEPRATIO	1.230	0.814
NOSO	1.200	0.832
SEX	1.170	0.852
Mean VIF	1.980	

Classification table

Classified	D	~D	Total
+	112	33	145
-	25	47	72
Total	137	80	217

Classified + if predicted Pr(D) >= .5	
True D defined as ADAPT != 0	
Sensitivity Pr(+ I D)	81.75%
Specificity Pr(- I ~D)	58.75%
Positive predictive value Pr(D I +)	77.24%
Negative predictive value Pr(~D I -)	65.28%
False + rate for true ~D Pr(+ I ~D)	41.25%
False - rate for true D Pr(- I D)	18.25%
False + rate for classified + Pr(~D I +)	22.76%
False - rate for classified - Pr(D I -)	34.72%
Correctly classified	73.27%

Appendix 10: Regression results considering 11 percent of the households alternatively as adapters and adapters.

The following tables show results of regression run by considering 11 percent of the households (27 in number) that performed adaptive strategies but did not report it as climate driven alternatively as adapters and non-adapters based on the discussion delivered in section 6.5 in chapter 6. Model 1 in the table assumes the households that performed adaptive strategies but simultaneously reported their behavioral response non-climate driven as non-adapters, whereas Model 2 assumes these households as adapters regardless of the reason (s).

1. Logistic regression results with perceptual variables.

Dependent Variable: Adaptive behavior (1= adapted; 0 = did not adapt)

VARIABLES		Model 1	Model 2
PV	3	1.245*	1.609**
		(0.742)	(0.664)
	4	2.605***	2.211***
		(0.749)	(0.682)
	5	2.911**	2.375*
		(1.337)	(1.351)
PS	3	-0.506	-0.859
		(0.679)	(0.766)
	4	0.415	-0.0392
		(0.627)	(0.728)
	5	2.103**	1.302
		(1.001)	(1.115)
AE	2	0.730	0.510
		(1.210)	(1.166)
	3	0.977	1.499
		(1.200)	(1.240)
	4	1.358	1.596
		(1.169)	(1.199)
	5	0.0407	1.317
		(1.192)	(1.232)
PC	2	-1.170	-0.0270
		(0.978)	(0.950)
	3	-0.0947	0.944
		(0.973)	(0.956)
	4	0.779	2.066**
		(0.954)	(0.959)

5	0.550 (1.083)	2.501** (1.259)
SE 2	2.551** (1.278)	1.253 (1.153)
3	1.537 (1.342)	-0.256 (1.322)
4	2.153 (1.334)	0.761 (1.309)
5	2.881* (1.501)	1.024 (1.488)
Constant	-5.102*** (1.415)	-3.948*** (1.334)
Linearity test	Significant _hat*** Insignificant _hatsq	Significant _hat*** Insignificant _hatsq
Likelihood ratio test	$\chi^2 = 78.25$*** (p=0.000)	$\chi^2 = 79.37$*** (p=0.000)
HL test	$\chi^2 = 13.76$* (p=0.0882)	$\chi^2 = 7.91$ (p=0.4423)
Correctly classified	76.2%	77.9%

2. Logistic regression results with socioeconomic variables.

Dependent Variable: Adaptive behavior (1= adapted; 0 = did not adapt)

VARIABLES	Model 1	Model 2
1.SEX	0.629 (0.409)	0.821* (0.439)
AGE	0.0190 (0.0267)	0.0114 (0.0295)
EDUC	0.0287 (0.0525)	-0.0104 (0.0570)
AGREXP	0.00821 (0.0252)	0.0260 (0.0278)
EXTN	0.0116 (0.0123)	0.0198 (0.0165)
1.ACCREDIT	0.414 (0.342)	0.880** (0.369)
NOSO	-0.314 (0.217)	-0.496** (0.238)
1.INFOWTH	0.718** (0.335)	0.757** (0.351)
LVST09	-0.0434 (0.0877)	-0.0214 (0.0958)
LANDSIZE	0.112 (0.238)	0.0219 (0.266)
DEPRATIO	-0.00101 (0.126)	-0.0773 (0.132)
DIST	-0.105 (0.232)	-0.247 (0.247)
Constant	-2.019** (0.924)	-1.545 (1.000)
Linearity test	Significant _hat*** Significant _hatsq**	Significant _hat*** Insignificant _hatsq
Likelihood ratio test	$\chi^2 = 30.65$** (p=0.002)	$\chi^2 = 59.33$*** (p=0.000)
HL test	$\chi^2 = 17.03$** (p=0.0298)	$\chi^2 = 12$ (p=0.1514)
Correctly classified	66.9%	74.3%

Model diagnostics proves Model 2 fulfilling all of the assumptions in both cases (psychological and SEI factors), but Model 1 fails to pass the Hosmer and Lemeshow model fitness test in both cases, and the test of linearity in the case of SEI variables. More importantly, the results in Model 2 exactly fit with the regression results reported in the main model (the model used for analyzing adaptive behavior in chapter 6). All of the significant and insignificant variables in the main model are similarly significant and insignificant in Model 2 with the exception of magnitude difference and sign dissimilarity in the case of 'education' as a variable. Such similarity of the results in this study implies that the variables that influence adaptive strategies regardless of the reasons may also influence climate driven adaptive behavior. The influence of the same variables in both cases further leads us to supplement our assumption of implicit learning of adaptation. As discussed in chapter 6, farmers might not even notice that they are adapting to climate variability and change while actually implementing some of the adaptation strategies due to the gradual and learned process of adaptation. Therefore, they may not consider it their adaptive behavior as climate driven even though the underlying reason may be climate.

Appendix 11: Household Survey Questionnaire.

The purpose of this survey questionnaire is to collect information on farming households as part of a field work for a PhD research.

I thank you in advance four your kind cooperation in answering the following questions and I would like to assure you that the information you offer will be analyzed anonymously and used for research purpose only.

I. General Information

Date ______________ Enumerator's name ____________________

☐ Region ____________ Wereda________________Kebele________

II. Household Information

A. Head of the household

☐ 1. Sex 1. Male 2. Female

☐ 2. Age: ________

☐ 3. Marital status 1. Married2. Single 3. Divorced 4. Widow/Widower

☐ 4. Farming experience in years: __________

☐ 5. Educational level: 1. Illiterate 2. Non-formal education 3A. Formal education (numbers of years of schooling):__

☐ 6. Religion 1. Orthodox 2. Islam 3. Protestant 4. Catholic 5. other ____

☐ 7. Ethnic group:____________________

B. Members of the household

8. Members of the household in age and sex (including the head)

	Age in years	Male	Sex	Total
8.1	9 & less than 9			
8.2	10-14			
8.3	15-64			
8.4	65 & greater			
8.5	Total			

9. Are there household members who participate in non-farm activities? If yes, how many members participate in non-farm activity?________________

10. How many household members are literate (read and write)?________________

III. Extension services and access to credit and market

11. Do you have contact with extension agents? 1. Yes 2. No

12. If your answer is 'yes' for the above question (item no.11), the average number of your contact with the extension agents per year is:________________

13. Do you have access to credit whenever you require it? 1. Yes 2. No

14. The distance between your house and the nearest output market in hours is:___

15. The distance between your house and the nearest input market in hours is: ___

16. The distance between your house and the nearest vehicle road in hours is: ____

17. Do you have access to transportation (vehicle) to go to market (to sell outputs or purchase inputs)?

1. Yes 2. No

1V. Membership in social organizations

18. Are you or anyone in the household members in social organizations (such as cooperatives, women, youth, and producer's association)? 1. Yes 2. No

19. If your answer is 'yes' for the above question (item no.18), the total number of social organizations the household is a member is:________________

20. How many relatives do you have in your residence Kebele? ________________

21. How many neighbors do you have? _______

22. Do you get (have access to) weather information on continual basis? 1. Yes 2. No

23. From who or where do you get weather/climate information most of the time?

1. Radio2. Extension agents 3. Neighbors 4. Various social organizations
5. If any others, please write: ________________

V. Crop production & livestock

24. Please list the household's annual crop production in the following table?

Crop produced in quintal										
No.	Year (E.C)	Teff	Wheat	Barely	Sorgh um	Millet	Corn	Rice	Haricot bean	Total
24.1	2002									
24.2	2002	Land area used for the production in hectares								
24.3	2001									
24.4	2001	Land area used for the production in hectares								
24.5	2000									
24.6	2000	Land area used for the production in hectares								
24.7	1999									
24.8	1999	Land area used for the production in hectares								
24.9	1998									
24.10	1998	Land area used for the production in hectares								

25. Please list the household's livestock ownership in the following table?

No	Year (E.C)	Cattle	Sheep	Goat	Camel	Donkey	Horse	Mule
25.1	2002							
25.2	2001							
25.3	2000							
25.4	1999							
25.5	1998							

VI. Agricultural inputs

26. In producing the crops listed in question item no. 24, please indicate the time you usually spent in each of the activities listed in the table below?

Activity	Land preparation & tillage	Planting /Sowing	Weeding	Harvest ing	Threshing/ Winnowing	Average
Time taken per day						

27. Please list the household's annual fertilizer and seed usage in the following table in the production of crops listed in question item no. 24?

No	Year (E.C)	Fertilizer in kilogram	Seed in kilogram
27.1	2002		
27.2	2001		
27.3	2000		
27.4	1999		
27.5	1998		

VII. Income and annual grain consumption

28. The amount of grain crop consumption of the household in quintal is: ________

29. Income source (s) of the household is (you can give more than one answer, if necessary)?

1. Crop sale 2. Livestock sale 3. Engagement in non-farm activity
4. Aid from relatives 5. If any other, please state: ________________

30. Which is/are the major source (s) of the household's income from the source(s) you indicated above in question item 29?

1. Crop sale 2. Livestock sale 3. Engagement in non-farm activity 4. Aid from relatives 5. If any other, please state: _____

31. Please indicate the annual income you earn from the following sources?

	Source of income	Birr
31.1	Annual income from the sale of livestock	
31.2	Annual income from the sale of grain	
31.3	Annual income from non-farm activities	

VIII. Land holding, irrigation and various farming related activities

32. Farm land holding of the household in hectares is: ___________________

33. The relative slope of your farmland is: 1. Plain 2. Gently hilly 3. Hilly

34. The fertility level of your farmland is: 1. Fertile 2. Moderately fertile 3. Less fertile 4. Infertile

35. Does your household practice irrigated agriculture? 1. Yes 2. No

36. If your answer is 'yes' for the above question (item no. 36), how many hectares of land is covered by irrigation? ________________

37. Please indicate the status of the following activities whether you have already implemented or not? What is your status of application with regard to the following actions?

	Activity	Implemented	Not implemented
37.1	Using different crop varieties	**1**	**2**
37.2	Changing planting dates	**1**	**2**
37.3	Adopting drought resistant crops	**1**	**2**
37.4	Increased use of soil and water conservation techniques	**1**	**2**
37.5	Diversification into non-farming activities	**1**	**2**
37.6	Water harvesting	**1**	**2**
37.7	Planting trees	**1**	**2**
37.8	Increased use of irrigation	**1**	**2**
37.9	Increased use of fertilizer	**1**	**2**

X. Malaria condition

38. Have you ever been infected by malaria? 1. Yes 2. No

39. Besides you, is there any member of the household who have been infected by malaria? 1. Yes 2. No

40. If you or any members of the household have been infected with malaria, please indicate the average frequency of sickness in the months listed in the table below? If there is no sickness, please fill zero or dash.

Sept. & Oct.	Nov. & Dec.	Jan. & Feb.	Mar. & Apr.	May & Jun.	Jul. & Aug.	Total

41. When you or a member of the household gets sick of malaria, the average number of sickness in days is: ________________

41.1 In the period of malaria sickness, how long do you stay (in number of days) in recovery without going to farm activities or work? ____________

41.2. (Asked based on the day difference given in question items 41 & 41.1). If you go to work before full recovery, how long hours do you work in the farms in these specific days of sickness?________________

42. Please list members of the household who have been infected by malaria in age and sex?

	Age	Male	Female	Total
42.1	9 & less			
42.2	10-14			
42.3	15-64			
42.4	65 & greater			
42.5	Total			

43. In the period of malaria sickness, usually who takes the responsibility of caring giving for the sick?

1. The household wife 2. The father 3. Daughter of the house
4. Son of the house 5. If other, please indicate: ________________

43.1. Age of the care giver is: ____________________

44. How many hours per day does the care giver spend on care giving (for sick) without going to work (farming)? _________

45. Please indicate the average expenses your household encounter at one time of malaria sickness?

Transportation (to go to the health center)	Medication	Other related costs (food, drink etc)	Total

46. Is there any history of death due to malaria in the household? If yes, indicate the number of deaths: ___________________

IX. Experience of risks, food aid and uncertainties (possible stressor factors in farming)

☐ 47. In the past 20 years, has your household been affected by the following climatic conditions?

	climatic condition	(Yes) ✓
1	Drought	
2	Flood	
3	Shortage of rainfall	
4	Failure in the timing of rain	
5	Heavy rain	
6	If any other:	

☐ 48. Have you ever encountered crop failure (partial or total failure)? 1. Yes 2. No

☐ 49. If your answer is 'yes' for the above question (item no. 48), how many times have you encountered the failure in the last twenty years? ____________

☐ 50. What is the reason for the crop failure (follow up question to question item no. 49)? ____

____

____

☐ 51. For how many months does your own farm production currently last to cover the food requirements of your family?

1. ≤ 3   2. 4-6   3. 7-9   4. 10-11   5. ≥ 12

☐ 52. If your answer for the above question is (item no. 51) 1 to 4, what is the major reason(s) for not to be able to cover the annual food requirements of the household)? ____

____

____

☐ 53. Do you receive food aid from government of non-governmental organizations?
1. Yes   2. No

☐ 54. If you get food aid, for how many years have you been on aid?____________

55. How do you rate the following factors as sources of stress to you? Please rate them by the scale indicated in the table below?

	Source of stress (uncertainty) factor	Not a source of concern at all (1)	Limited source of concern (2)	Moderately a source of concern (3)	Highly a source of concern (4)	Very highly a source of concern (5)
55.1	Lack of agricultural inputs (fertilizer, seed etc)	1	2	3	4	5
55.2	High price of inputs	1	2	3	4	5
55.3	Lack of vehicle road	1	2	3	4	5
55.4	Lack of market to purchase inputs and sell outputs	1	2	3	4	5
55.5	Lack of credit	1	2	3	4	5
55.6	Lack of extension services	1	2	3	4	5
55.7	Drought	1	2	3	4	5
55.8	Flood	1	2	3	4	5
55.9	Insufficiency of rain in farming season	1	2	3	4	5
55.10	Irregular and unpredictable rains	1	2	3	4	5
55.11	Occurrence of heavy rain in few days	1	2	3	4	5
55.12	Delay of rains in farming season (late onset)	1	2	3	4	5
55.13	Too early rainfall which is not sustainable	1	2	3	4	5
55.14	Early cessation of rain in farming season	1	2	3	4	5
55.15	Warmness of days	1	2	3	4	5
55.16	Malaria	1	2	3	4	5
55.17	Lack of oxen	1	2	3	4	5
55.18	Limited availability of farm land	1	2	3	4	5
55.19	Low level of soil fertility	1	2	3	4	5
55.20	Population pressure	1	2	3	4	5
55.21	If any other, state & rate					

XI. Perception on local climate

56. How do you feel the trend of following weather or climatic conditions in the last twenty years in your area?

56.1. The amount of precipitation: 1. Decreasing 2. Increasing 3. No change 4. I do not know

56.2. Temperature: 1. Decreasing 2. Increasing 3. No change 4. I do not know

57. Is there any adaptation mechanism or strategy in agricultural sector you carry out as a response to climatic problems you usually experience in the locality or as a response to rainfall and temperature increasing/decreasing trends you indicated above? 1. Yes 2. No

58. If your answer is 'yes' for the above question (item no. 57), would you (if possible) classify the adaptation mechanisms you carry out whether they are meant to adapt specifically to rainfall problem or temperature problem?

Adaptation strategies as a response to rainfall problem	Adaptation strategies as a response to temperature problem

59. What barriers or difficulties do you encounter in you effort towards implementing adaptation strategies?______________________________

____

____

60. Based on your experience and observation, how do you feel about the level of susceptibility (vulnerability) of your crop farming to usually experienced forms of climatic problems and climatic conditions you indicated above (rainfall and temperature increasing/decreasing trend)?[90]

Not vulnerable at all	Less vulnerable	Moderately vulnerable	Highly vulnerable	Extremely vulnerable
1	2	3	4	5

[90] For those who have undertaken adaptation measures, additional explanation was given to the respondents to answer the question with the assumption that if they did not have implemented the measures.

61. Based on your experience and observation, how do you feel about the level of harm or damage that your crop farming could face being exposed to usually experienced forms of climatic problems and climatic conditions you indicated above (rainfall and temperature increasing/decreasing trend)?[91]

Not harmful at all	Less harmful	Moderately harmful	Highly harmful	Extremely harmful
1	2	3	4	5

62. How do you rate the effectiveness of the following actions in helping to reduce the risks in crop farming from problems associated with climatic variability and change or unfavorable climatic conditions that you usual experience?

	Activity	Ineffective (1)	Somewhat ineffective (2)	Neither ineffective nor effective (3)	Somewhat effective (4)	Effective (5)
62.1	Using different crop varieties	1	2	3	4	5
62.2	Changing planting dates	1	2	3	4	5
62.3	Adopting drought resistant crops	1	2	3	4	5
62.4	Increased use of soil and water conservation techniques	1	2	3	4	5
62.5	Diversification into non-farming activities	1	2	3	4	5
62.6	Water harvesting	1	2	3	4	5
62.7	Planting trees	1	2	3	4	5
62.8	Increased use of irrigation	1	2	3	4	5
62.9	Increased use of fertilizer	1	2	3	4	5

[91] The same assumption in footnote number 1 applies as well.

63. How confident do you feel in your household's resource capacity (financial capacity) to be able to perform the following actions? Please rate your level of confidence?

	Activity	Unconfident (1)	Somewhat unconfident (2)	Neither ineffective nor effective (3)	Somewhat confident (4)	Confident (5)
63.1	Using different crop varieties	1	2	3	4	5
63.2	Changing planting dates	1	2	3	4	5
63.3	Adopting drought resistant crops	1	2	3	4	5
63.4	Increased use of soil and water conservation techniques	1	2	3	4	5
63.5	Diversification into non-farming activities	1	2	3	4	5
63.6	Water harvesting	1	2	3	4	5
63.7	Planting trees	1	2	3	4	5
63.8	Increased use of irrigation	1	2	3	4	5
63.9	Increased use of fertilizer	1	2	3	4	5

64. How confident do you feel in your skill or expertise to perform the following actions? Please rate your level of confidence?

	Activity	Unconfident (1)	Somewhat unconfident (2)	Neither ineffective nor effective (3)	Somewhat confident (4)	Confident (5)
64.1	Using different crop varieties	1	2	3	4	5
64.2	Changing planting dates	1	2	3	4	5
64.3	Adopting drought resistant crops	1	2	3	4	5
64.4	Increased use of soil and water conservation techniques	1	2	3	4	5
64.5	Diversification into non-farming activities	1	2	3	4	5
64.6	Water harvesting	1	2	3	4	5
64.7	Planting trees	1	2	3	4	5
64.8	Increased use of irrigation	1	2	3	4	5
64.9	Increased use of fertilizer	1	2	3	4	5

Thank you once again for giving me the information by allotting your precious time.

Bisher im Logos Verlag Berlin erschienene Bände der Reihe

UAMR Studies on Development and Global Governance

ISSN: 2194-167X

Vormals "Bochum Studies in International Development"
(ISSN: 1869-084X, Vol. 56-60)

56	Martina Shakya	Risk, Vulnerability and Tourism in Developing Countries: The Case of Nepal ISBN 978-3-8325-2275-9 47.50 EUR
57	Tobias Birkendorf	Die Auswirkungen von Großveranstaltungen auf die langfristige öknomische Entwicklung von Schwellenländern. Eine wachstumstheoretische und empirische Untersuchung am Beispiel der Olympischen Sommerspiele in Beijing 2008 ISBN 978-3-8325-2285-8 41.00 EUR
58	Andreas Hahn	Between Presidential Power and Legislative Veto. The Impact of Polity and Politics on Economic Reforms in Brazil ISBN 978-3-8325-3340-3 38.50 EUR
59	Kai Riewe	Interdependenz wirtschaftlicher und gesellschaftlicher Transformation. Das ökonomische Potential sozialen Kapitals am Beispiel der Russischen Föderation ISBN 978-3-8325-2598-9 40.00 EUR
60	Tobias Bidlingmaier	The Influence of International Trade on Economic Growth and Distribution in Developing Countries. With a Special Focus on Thailand ISBN 978-3-8325-2621-4 35.00 EUR
...	....	....
61	Shafaq Hussain	Growth Effects and the Determinants of Female Employment in Pakistan: A Macro- and Microeconomic Analysis ISBN 978-3-8325-3131-7 37.00 EUR

62	Abate Mekuriaw Bizuneh	Climate Variability and Change in the Rift Valley and Blue Nile Basin, Ethiopia: Local Knowledge, Impacts and Adaptation ISBN 978-3-8325-3524-7 42.50 EUR

Alle erschienenen Bücher können unter der angegebenen ISBN-Nummer direkt online (http://www.logos-verlag.de) oder per Fax (030 - 42 85 10 92) beim Logos Verlag Berlin bestellt werden.